Body, Soul, and Bioethics

Body,
Soul,
and
Bioethics

GILBERT C. MEILAENDER

University of Notre Dame Press
Notre Dame and London

Library of Congress Cataloging-in-Publication-Data

Meilaender, Gilbert, 1946–
    Body, soul, and bioethics / by Gilbert Meilaender.
        p.   cm.
    ISBN 0-268-00698-9 (hc : alk. paper)
    1. Medical ethics.   2. Bioethics.   I. Title.
    R724.M35   1995
    174'.9574—dc20   95-17486
                                                                CIP

∞ *The paper used in this publication meets the minimum requirements of the*
American National Standard for Information Sciences—Permanence of
Paper for Printed Library Materials, ANSI Z39.48-1984.

In memory of

Paul Ramsey,

who saw deeply into the truth
of these matters

Contents

Preface

Within the community of those who think and write about bioethics there is an increasing self-consciousness. What began as a movement some twenty-five to thirty years ago now seems to have become almost a discipline. Self-consciousness about "method" is one sign of that shift, as is a constantly expanding body of literature about central concepts such as autonomy and personhood.

The five chapters of this book are intended as a contribution to that self-consciousness, although they are no doubt rather unconventional. If forced to describe myself, I would claim to be a theological ethicist, not a bioethicist. I hope, of course, that these chapters are not uninformed about some of the bioethics literature, but, nevertheless, my angle of vision is not that of one who exists chiefly in the bioethics world. I can only hope that such a perspective has, in addition to its weaknesses, some insight to offer.

I take up here only a small range of topics within bioethics, but these topics permit me to suggest that the development of bioethics over the past quarter century has not been entirely benign. An increasing focus on public policy often obscures the importance of background beliefs about human nature and destiny. Without drawing attention to those beliefs we cannot always fully see what is at stake in many of our debates.

Chapters 1 and 4 below focus, in quite different ways, on how easily the "soul"—attention to the meaning of being human, a meaning often illuminated by religious and metaphysical insight—can be lost in bioethics. Chapters 2 and 3 explore problems at the end and the beginning of life in an attempt to uncover some of those deeper issues that need our attention. Finally, chapter 5 takes up briefly abortion, the issue that so often lurks just beneath the surface of bioethical argument. (Earlier versions of chapters 2 and 5 appeared in the *Hastings Center Report*, and an earlier version of chapter 3 appeared in the *Washington and Lee Law Review*.) Throughout I attempt to emphasize the "soul" of these issues—questions about who we are and what we may become. And throughout I suggest that recapturing that soul will lead us to a new appreciation of the living body as the locus of personal presence.

1. How Bioethics Lost the Soul: Questions of Method

Bioethicists have begun to reflect upon the history of their discipline. Whether that is a sign of maturity or unease may be debated, but it is a fact. Albert Jonsen dates "the birth of bioethics" from the year 1962, when Shana Alexander's article describing the Seattle dialysis selection committee appeared in *Life* magazine.[1] Elsewhere Jonsen describes 1965–75 as the "formative decade" for bioethics in this country.[2] David Rothman, in what is the first history of the bioethics movement, dates its beginning with the 1966 publication of Henry Beecher's article exposing abuses in human experimentation. Less concerned than Jonsen to focus on bioethics alone, Rothman attends more to the great changes in American medicine that were taking place—changes in which bioethics played an important role. For him the "critical period of change" was 1966 to 1976, beginning with Beecher's article and ending with the Quinlan decision by the New Jersey Supreme Court. During this decade physicians increasingly became "strangers" to their patients, and, simultaneously, a new set of strangers—bioethicists—established their role as authority figures near the bedside.[3] Thus, despite somewhat different concerns, Jonsen and Rothman agree that the bioethics movement is now better

Lles = start

than a quarter century old. It is not surprising that the time has come to take stock.

Not everyone is happy with the current state of bioethics, and many of the discontents focus on questions of method. From where should bioethics take its direction? From within the practice of medicine itself, or from more general moral norms applied to medicine? Do we need a moral theory to guide our bioethical reflection, or can we make our way from case to case, gradually mapping the territory? Are the most common bioethical approaches focused too much on the language of rights, able to offer only a thin and minimal ethic that gives little real wisdom about how we ought to live and die? Clearly, these questions about method in ethics have been shaped in large measure by substantive issues in medical practice. Shall physicians reveal to their patients only as much truth as they think wise? Or do patients have a right to know as much as the physician can explain about their circumstances? If an ethic generated entirely from within the practice of medicine too easily answered yes to the first of these questions, the claim that bioethics involves the application of general norms to medicine was designed to answer yes to the second question. Questions of method and of substance have been intertwined in debates about the nature of bioethics

In exploring debates about method in bioethics, however, I intend to develop gradually a certain vision of what has been happening over the quarter century in which bioethics has risen to something approaching the status of a discipline. Bio-ethics has, over this period, lost the soul. Our soul. By this I mean that it has to some considerable degree turned away from exploration of the most fundamental questions about who we are and should be. No method, no theory, operates apart from a collection of background beliefs that inform the theory's use, but it is possible to ignore those background beliefs and suppose that we can "do" bioethics without uncovering or unpacking them. Such attempts are, I think, ultimately futile. It would be a drastic overstatement, of course, to suggest that the description I have just given needs no qualification or fits every bioethicist. Nevertheless, this chapter's discussion of some of the most important methods currently used may indicate how easily deeper metaphysical and religious concerns can be and often have been lost.

I

If medicine is a profession, physicians must have something to "profess." That is, they cannot simply be skilled technicians who place their abilities at the service of patients' desires, as if their profession itself had no larger meaning of its own. And, in fact, even though the rise of bioethics was due in part to a perceived arrogance of traditional medical practice, I doubt that we really want physicans to understand themselves simply as such technicians. We want a profession with something to profess, with the ability to maintain a certain critical distance from the larger society.[4] Indeed, the ancient professions such as medicine and law—like religious communities and like the family—give a person a standing ground over against civil society. The good person, holding membership in such mediating structures, cannot simply be presumed equivalent to the good citizen.

The Hippocratic tradition pictured medicine as a profession with its own internal goals and norms, providing "the necessary and sufficient ingredients for a coherent medical ethic from within the culture of medicine itself."[5] From this perspective—held by contemporary thinkers such as Edmund Pellegrino and Leon Kass—the norms of medicine are generated from within the practice itself; they are not the "application" to medicine of more universal norms.

We are sometimes so impressed with the diversity of our society that we do not take this view as seriously as it deserves to
be taken. Thus, for example, Kass argues that doctors must not
kill, that to kill patients violates the essential meaning of the
practice of medicine.[6] It will not suffice to respond by noting
that some physicians disagree. That there will always be such
disagreements within a profession seems likely, but it is not ter-
ribly significant. A profession does not consist simply of the
sum total of its members, nor does the definition of virtuous
professional practice depend solely on the opinions of current
practitioners. There will always be some gap between the opinions of individual practitioners and the purposes said to be
inherent in the profession itself. Of course, if that gap becomes
too large, we are rightly concerned. If the gap persists and
widens, we might eventually have to conclude that the very
meaning of the profession has to be reconceived.[7] But that
reconceptualization would require normative argument, not
surveys of current opinion. And we might conclude that what
was really needed was not reconceptualization but better
initiation and socialization into the norms of professional practice—better moral education of beginning practitioners. To
suppose that the visions of professional virtue could be as many
as there are practitioners would be tantamount to giving up the
notion of a profession as a social institution. Thus, even in a diverse society there may be norms inherent in a profession.

To think of medicine's goals and norms as internal to its
practice is not a rejection of moral theory; it is the adoption of
an Aristotelian understanding of morality. On this view, appropriate behavior is not determined by the reflective application
of moral rules, as if we determined what we ought to do by
specifying more precisely the application of universal rules to
particular cases. Instead, appropriate behavior is learned as one
gradually learns a way of life—the habits of conduct that constitute its grammar and syntax. The model is closer to that of

apprenticeship than the classroom, and it is not accidental that still today much of what is most important in medical education adheres to such a model.

From this perspective, therefore, physicians know at least some of their duties entirely apart from more general moral principles. They know they ought not deceive their patients or have sexual relations with them not simply because such activity violates a universal moral rule but because it fails to take seriously the vulnerability of patients before physicians, the asymmetry and inequality of the relation. They know they ought not kill their patients (for love or money, as Kass puts it) not because they adopt a general rule against killing. Indeed, as citizens they might well serve as soldiers. But they ought not use their skill as physicians to serve, for example, as state executioners (via lethal injection), even though they need have no principled objection to capital punishment. Not more general moral claims but the moral meaning of the profession itself tells physicians that they must be committed to the bodily life of their patients, whose personhood "is manifest on earth only in living bodies."[8]

The history of the past quarter century of bioethics is in large measure a story of movement away from such a view. It would be an exaggeration, but not a complete misrepresentation, to say that bioethics developed over against a view of traditional medical ethics that was perceived as paternalistic, arrogant, and elitist.[9] Daniel Callahan has described this as a movement from an "internalist" to an "externalist" view of the relation of medical morality to the wider moral standards of society.[10] For the internalist view, medical morality is largely self-contained, its principles internal to the practice of medicine. The externalist view, by contrast, thinks that the ends of medicine must be determined from without, by the morality of the larger society. And the physician's skills are then placed in service of those ends. "On the whole," Callahan writes, "I believe we have witnessed over the past few decades a movement from the Hippocratic, internalist view, to the externalist perspective."[11]

To see this shift is to understand why the moral norm given pride of place within bioethics has most often been patient autonomy. That emphasis is the substantive result of a shift in method. Logically, the substantive and methodological shifts could be separated (since the general norms of a society might specifically countenance medical paternalism), but, clearly, they go well together. Nor is it surprising, therefore, that bioethicists—for example, Robert Veatch, Tristram Engelhardt, Tom Beauchamp, and James Childress (forever joined by a coordinating conjunction)—who are vigorous proponents of patient autonomy should also be robust critics of what Callahan calls the "internalist" perspective.

Thus Veatch argues that "[s]pecial norms . . . cannot exist for a professional group without collapsing into ethical relativism and particularism."[12] If there could be circumstances in which it would be "wrong" for a doctor to kill a suffering person but "right" for someone else, then, so the argument goes, professional ethics would have become an ingrown, relativist code. I do not myself think this is the best way to put the problem, though, to be sure, Veatch is directing our attention to something important. The problem is not precisely relativism. For, after all, on Kass's account, anyone—past, present, or future; at any time or place—professionally committed to the healer's role ought not kill those in his care. The problem, rather, is that such claims to special professional virtues and obligations may sometimes seem to stand in tension with larger social goods. If we permit this tension to stand, we would, in Larry Churchill's words, have to give up "the ideal of morality as necessarily unitary in nature."[13] The problem, then, is not relativism, but the very old question of the unity of the virtues. Will good physicians necessarily be good citizens?

Veatch is, in effect, arguing that they must be, that the virtues of physician and citizen must be harmonized by deriving the former from the latter, that the task of bioethics is the application of general norms to the practice of medicine. We can understand and honor such an impulse toward unity and universality in the moral life. It may be too easy to stress the disunity of the virtues and the tensions of moral life, and we can pay a price for savoring the tragic too much—a price that compartmentalizes human character, settles for isolated virtues, forfeits the ability to talk about any way of life that is best for human beings, and loses the sense that to seek virtue is to seek wholeness of self.

We will, therefore, need to attend with care below to the externalist perspective, according to which bioethics applies more general norms to the practice of medicine. For the moment, however, we can consider a different alternative. If we grant (with Kass) that at least some norms of medical practice are generated internally, and if we grant (with Veatch) that these may sometimes stand in tension with more general norms, and if we do not wish to espouse a moral view (with Churchill) that accords finality to tragic conflict between our several obligations, we might be attracted by Ezekiel Emanuel's attempt to sketch a communitarian medical ethic. Beginning with the Aristotelian belief that there are moral norms internal to medical practice but that these cannot solve all of medicine's moral problems, Emanuel makes a standard Aristotelian move: Ethics is a branch of politics. "Medical ethics is . . . a subfield of political philosophy."[14] The norms of medical practice are not simply derived from a universal morality, but neither are they entirely separate. Nor can we be content to recognize different spheres of moral obligation always in tension.

The solution? An "integration" of medical ethics and universal morality within a community (or communities) in which there is agreement not just on a few general principles but on the meaning of the good life. Continued commitment to our liberal political tradition must inevitably, Emanuel argues, lead to frustrating deadlocks over the problems of bioethics. If the community as a whole need not share or promote any substantive vision of what is good for human beings, we cannot expect to agree about what constitutes proper medical care. We need, therefore, a federation of Community Health Programs (CHPs) in which groups of citizens thrash out together a shared understanding of the good life and let that understanding shape the practice of medicine within their CHP. This will, of course, require considerable commitment on our part to mutual conversation and discussion. Indeed, the time required may be so great that one might suspect we are paying away the very lives that medicine exists to enhance! We are not likely to do this unless we are persuaded that a liberal polity is truly unworkable.

As an example of his claim that our defective political philosophy makes arguments about medical care insoluble, Emanuel takes up the problem of distribution of medical resources. He argues that we cannot arrive at shared principles of distribution, because any such principle will depend upon a vision of the good life for human beings—about which, he says, we must be agnostic in a liberal polity. He notes that we could, of course, simply grant that any of several possible distributive schemes would be just and then let democratic political procedures (i.e., votes) determine which of these procedures to adopt.

In Emanuel's view, however, such an approach cannot work. On what basis are we to cast our votes? If we ask which distribution scheme is more just, the answer can only be that all pass the threshold of justice. If we ask which most enhances the good life for human beings, we have, according to Emanuel, asked a prohibited question. Hence, we can only vote in accord with our interests—evidently, in his view, a debased understanding of communal life.

It is worth pausing over each of these possibilities—that we might vote in accord with our interests, and that we might vote in accord with our vision of the good life. Because he regards neither of these possibilities as desirable, Emanuel does not want our disagreements to "devolve into voting."[15] He prefers a consensus worked out in public discussion and debate. I confess that the preference for consensus has little charm for me. To accept the results of a vote is one way of sharing a common life—a way which, incidentally, leaves us free to get on with the rest of life. To lose a vote is simply that and nothing more. It does not require that we pretend to think the majority decision wise. By contrast, decision by consensus often carries a subtly coercive quality. When we acquiesce in a consensus that does not fully persuade us, our acquiescence is often interpreted, not simply as acceptance of democratic voting procedures, but as approval of a decision we may in fact have opposed.

Would it be better just to vote our interests and see where majority opinion lies? In many instances it might be. We can think of such votes as "a form of ritual combat that provides not only for the decision of political issues but above all for a nonviolent transfer of power."[16] In accepting voting patterns grounded in self-interest, we are not returning to a Hobbesian state of nature. Just as we may be confident that our fellow citizens will, like us, have interests, so may we be confident that, like us, they will have some sense of fairness and some willingness to think in terms of what is best for the community as a whole.[17]

Emanuel's other alternative to a search for consensus is an attempt to persuade others of the truth of our vision of what is best for human beings. And, of course, visions of the good life will conflict every bit as much as interests do. We need not bemoan all such conflict.

The meeting of will with will, which almost always involves conflict at some level, is the very substance of personal relationships, which would not be fully personal without it. It is through conflict, or something like it, that we know the otherness of self and other. The oppositions that arise between our wills and our parents' wills are necessary for our differentiating ourselves from them. The fact that the world is in some ways contrary, and in some ways unresponsive, to our wills is what keeps us from regarding it all as an extension of ourselves. . . . We have learned to be suspicious of marriages in which the spouses claim they have no fights. In politics, likewise, conflict could hardly be eradicated without excluding from the political process the selfhood of most of the individuals, and the identity of many of the groups, in the society.[18]

Emanuel too quickly assumes that we have acted illicitly or illiberally if we make decisions, formulate arguments, and cast votes guided by our vision of the good life. In so doing, we do not seek to "impose" that view; we seek only to persuade. Understanding that what persuades us may not always persuade others—understanding, that is, the limits of moral reasoning—we agree to let that "form of ritual combat" that is voting be determinative (at least for now). Of course, if every component of my vision of the good life must become an absolute right (constitutionalized, perhaps, through an expanded right of privacy), and if the same must be true of every component of your vision of the good life, then certainly Emanuel's belief that a liberal polity is unworkable will be correct. But that is a form of individualism that cannot adequately understand the individual as one who is always situated within given communities, and it is only a different way of retreating from the conflicts that are the substance of personal relationships.

We need not and should not accept such a description of a liberal polity. Indeed, were thoughtful persons able to place themselves not behind the "veil of ignorance" described by John Rawls but into a genuinely neutral "original position," they would not, I believe, find Emanuel's depiction of a liberal polity choiceworthy. For they would be asking others to withhold from them the kind of education and argument that could have a profound effect upon their understanding of how they ought to live.[19]

We are, therefore, free to continue to ponder the course bioethics should take without a radical restructuring of our political community. For one way to characterize Emanuel's project is to see in it a heroic attempt to reverse the rise of bioethics. Emanuel does not want bioethicists at the bedside; he wants each physician to undertake this task of moral integration, each physician as his own bioethicist. "The aim of medical ethics, then, should not be to create a pool of bioethicists for physicians to consult for the resolution of dilemmas, but to make physicians recognize the intrinsic ethical nature of their practice and to understand the relationship between the ends of medicine and political philosophy."[20] This extraordinary exaltation of the physician as savior figure will, of course, resonate with much that is deeply embedded in our tradition and our psyches, but it ought to be resisted. If we need not devote ourselves to endless discussion in CHPs in order to live a full life, we also need not relinquish our claim to enter whenever we wish into bioethical debate.

To be sure, moral theory will not save us either. But to suppose, as Emanuel does, that we are wrong even to search for such universal principles is, I think, a misplaced description of our problems. If there turns out to be tension between the general moral principles we all bring to medicine and the internally generated norms of medical practice itself, we may have to live with that tension. Indeed, such tension within moral theory reflects substantive truths about medical practice. In the last quarter century bioethics has moved increasingly away from a paternalism thought intrinsic to the doctor's role and toward an application of a general principle of autonomy to the physician/patient relation. Yet, the truth may well be that the interplay of patient and physician is substantively too complex for either of these descriptions to be adequate. Given the facts of patient vulnerability and physician expertise, a patient's autonomy cannot be achieved autonomously. It can be accomplished only with the physician's help.[21]

Emanuel's attempt to overcome these complexities and unify the moral life almost inevitably overstates the claims of the political realm. Of course, standard attempts to understand bioethics as simply the application to medicine of external, more general moral principles are also attempts to unify morality. Such "applied ethics" approaches have dominated bioethics, but they too have their problems and their critics.

II

Perhaps the very best example of the "applied ethics" model for bioethics is *Principles of Biomedical Ethics*, by Tom Beauchamp and James Childress.[22] The Beauchamp/Childress approach has even been given a name—dubbed "principlism" by its critics. And a recent volume, devoted largely to critiques of this method, has described principlism as a "patient" who is "ill."[23]

Beauchamp and Childress are clear in their insistence that the norms of bioethics are not internal to the practice of medicine. To be sure, professional practice necessarily has a kind of priority. "Ethical theory does not create the morality that guides professionals' decisions and actions. It can only cast light on and supplement that morality, in part by analyzing and appraising moral justifications, their presuppositions, and their implications."[24] But the moral framework they provide for work in bioethics is based upon a belief that "standards developed by the medical profession have suffered from the absence of an external basis for the justification and revision of these standards" (p. 21). While arguing that bioethics must have a grounding external to medical practice itself, Beauchamp and Childress do not mean to defend what is now called moral "foundationalism," the view that our moral knowledge is based not upon history and tradition but upon ahistorical, universal truths. We might describe their view of morality, in John Reeder's words, as "foundations without foundationalism."[25] Morality, according to them, is rooted in communities, but these communities are larger than the body of medical practitioners alone.

Usually, however, antifoundationalist theories of morality attack not only the idea of ahistorical foundations but also a picture of moral reasoning as deductive (i.e., that we reason from those universally justified principles to more specific judgments). Antifoundationalists are more likely to describe moral reasoning in terms of a movement back and forth between the general principles we hold and our judgments about specific cases. In this dialectical movement we seek coherence between principles and particular judgments, a coherence that can be gained by adjusting either the principles or the judgments.[26] It is, therefore, a little surprising that Beauchamp and Childress should picture justification in moral reasoning as straightforwardly deductive. They describe (complete even with diagram) a process by which particular moral *judgments* are justified by appeal to moral *rules*, which are in turn justified by more general *principles*, which themselves are justified by an ethical *theory* (pp. 6f.).

Beauchamp and Childress are not, in fact, tied to a rigidly deductive model of moral reasoning, but the heuristic clarity of this approach has surely accounted in part for its popularity and influence.[27] It is useful for teaching, and its categories for analysis may also have appealed to practitioners seeking terms within which to conceptualize their problems. Beauchamp and Childress did not, of course, invent this approach to bioethics. Its initial formulation came in the work (from 1975–78) of the

National Commission for the Protection of Human Subjects of Biomedical and Behavioral Research. The Commission's *Belmont Report*, in which it sought briefly to articulate the method underlying its deliberations on a variety of topics, made three principles—respect for persons, beneficence, and justice—central.[28] Beauchamp and Childress turned these three principles into four that became the heart of their method: respect for autonomy, nonmaleficence, beneficence, and justice.

A peculiarity of their method, however, is that theory often turns out to seem "idle." For example, when one applies their principles one often encounters tensions between several of the principles—especially, perhaps, between respect for autonomy and beneficence. Since for Beauchamp and Childress, the principles themselves are only prima facie binding, such tensions seem like perfect moments to move to the higher level of theory in search of a resolution. One might suppose that a utilitarian theory would tilt us in the direction of beneficence, while a deontological theory would lend support to autonomy when the principles were in conflict. Beauchamp and Childress themselves differ at the level of theory—one holding that rule utilitarianism is the more satisfactory theory, the other that rule deontology is better. Yet, these theoretical differences seem to make little difference in their work. Indeed, they write that the differences between the two theories can be "exaggerated" and that often these different theoretical commitments will lead to "virtually identical principles and rules and recommended actions" (p. 44). Why then, the reader may wonder, should these competing theories play such an important role in the sketch of their method?

Nor does this end the bewilderment. Having granted that the differences between moral theories "are not as significant . . . as they have sometimes been taken to be" (p. 46), they adopt the view that the four basic principles they have developed are "binding but not absolutely binding" (p. 51). When particular judgments are needed about cases, we can only "weigh" their respective claims—a procedure for which the theory itself gives us no guidance.[29] And this, in turn, forces them to admit that the differences between their theoretical commitments and those of situational theories (act-utilitarianism or act-deontology) may in practice be "minimal" (p. 53). Hence, they write, "[j]ust as we earlier argued that the consequentialist/deontological distinction may not be as significant for moral theory as it has sometimes been taken to be, so we might reasonably conclude now that the distinction between act and rule theories can, without careful analysis, create more obscurity than insight" (p. 54). Here again discussions of theory seem largely idle.

Ultimately, however, it is more important to see that the theory is "thin" than that it is sometimes idle. *Principles of Biomedical Ethics* often fails to provide the kind of wisdom we need most. Beauchamp and Childress acknowledge that application of their principles will "depend on factual beliefs about the world" (p. 7). How we describe a situation will depend on the background beliefs—scientific, metaphysical, and religious—that we bring to it. They recognize the relevance of such beliefs but are seldom willing to explore them in detail. Several examples will serve to illustrate this unwillingness.

In their chapter on the principle of nonmaleficence, Beauchamp and Childress address the issue of care for incompetent patients. Their general tendency is to begin with an obligation to treat, but that obligation may, they believe, be overridden in some situations. Having developed their argument, they grant, however, that "there is an important reason why this argument from nonmaleficence . . . remains incomplete" (p. 162). Something important has been left undone. We need to know

not only the applicable rules and their priority when they conflict but also to whom these rules apply. If we are certain that a particular being is a person, then we can usually be confident that the full complement of moral principles applies. But if, for example, we make a judgment that fetuses or certain neonates are not persons or are not properly the objects of rules about sustaining life . . . , this verdict would dramatically alter the requirements of newborn care. . . . This controversy about human life and personhood is impossible to decide on the basis of the moral principles that form the core principles in this book. (P. 162)

Yet, of course, the unanswered questions are ones about which we desperately need wisdom and guidance.

Similarly, in their chapter on beneficence, Beauchamp and Childress consider whether there are any limits we can set to this obligation. Do we have to benefit others even if doing so will make us worse off? Some have argued in ways that suggest we are obligated to exercise beneficence toward others until we reach the point that we are in the same position as "that of the world's most destitute person" (p. 199)—until, for all practical purposes, we have nothing left to give. To this Beauchamp and Childress respond quite sensibly that such an understanding of the obligation of beneficence seems "overly demanding. Normal moral conventions establish limits to our obligations to help others . . ." (p. 199). True enough, but we would like to know why. An adequate answer might have to explore some disputed beliefs about human nature—whether we have special role responsibilities, whether our own self-fulfillment should be of moral importance to us, whether we might have a "calling" in life that is not reducible to an obligation of general beneficence, whether God might join our several vocations in a "cosmos of callings" that would be generally beneficent even if we are not obligated to reduce ourselves to destitution in the name of beneficence.[30] In their discussion of beneficence, Beauchamp and Childress also express the belief that "both individuals and their family members have an obligation of beneficence, to donate cadaver organs to benefit patients suffering from end-stage organ failure" (p. 207). At the same time, however, they hold that "society should not enforce this obligation of beneficence against the decedent's prior wishes or the family's wishes, in order to respect autonomy and avoid offense and outrage" (p. 207). We get little clue, though, about whatever process of weighing there is that would lead us here to give preference to autonomy over beneficence. Perhaps it involves the sense that our bodies are the locus of our personal presence, that it is unseemly to force a family to fight for the body of a loved one at the very time when it is trying to ritualize the loss of that loved one.[31] Once again, the seeming clarity of the theoretical structure, which I do not wish to underrate, encourages us to forget (what Beauchamp and Childress themselves do not forget) that many of the most important questions are being left unanswered.

One more example of the thinness of the theory will suffice. When they turn, in particular, to the doctor-patient relation, Beauchamp and Childress have occasion to move from their four basic principles to rules (about truthtelling, fidelity, privacy, etc.) derived from those principles. They believe that a patient's right to privacy is grounded chiefly in the principle of respect for autonomy (p. 321). An obvious difficulty, however, is that a patient who has never been autonomous and has no future capacity for autonomy would then seem to lack such rights as "the rights not to be needlessly viewed or touched by others" (p. 322). And, as Beauchamp and Childress sensibly suggest, it is likely to seem to us "intuitively correct to say that it is a violation of privacy to leave a comatose person undraped on a cart in the hospital corridor, not merely a tasteless act of negligence" (p. 322). However, having for the most part eliminated from their discussion any detailed description of the nature of the persons possessing such rights, they can say little more in support of such an intuition. They can only note the possibility, which they neither pursue nor defend, that respect for persons might not be grounded in autonomy alone but in some more general sense of "human dignity" (p. 322). Were that notion of human dignity unpacked, however, it might begin to make a difference in what we think about many of the issues taken up in *Principles of Biomedical Ethics*. But then, I suspect, the theory might not accomplish their aims.

For this is a theory fashioned largely with public policy in mind, driven by the search for consensus in a pluralistic society. The "principlist" method seeks to fashion a minimal morality for a community of strangers, even if friendly strangers. We ought not underrate this; for, as Childress has recently noted, at least in the world of medical care we may inhabit precisely such communities.

> Major changes in health care have rendered problematic a conception of medicine in terms of friendship. Pluralism in values; the decline of close, intimate contact over time between professionals and patients; the rise of specialists who treat only part of the whole person; and the growth of large, impersonal, and bureaucratic institutions of health care have all contributed to the loss of intimacy and community. . . . In the absence of community, then, principles, rules, and procedures become increasingly important.[32]

It remains true, however, that a bioethics fashioned for this purpose will offer a lowest common denominator agreement. It will bracket matters on which we might intensely disagree— the nature of the human person, the meaning of suffering, the foundation of human dignity. Its focus is public, and, aiming at consensus on policy, it is more likely to lead to moral routinization than to prophetic witness.[33]

We can begin to understand, then, how it is that the Beauchamp/Childress approach is often thought of not simply as a *method* for bioethics but as a defense of one particular principle, autonomy. Understood simply as a method of inquiry, "princi-plism" need not enshrine any particular normative principle. But as a method aimed at shaping public policy and consensus, it seeks to avoid ultimate contexts of meaning and visions of the good life that citizens may not share. Hence, the heavy weight placed upon autonomy—a weight required not by the theory's structure but by its proponents' aims.[34] If we can agree on nothing else, we may at least agree to respect each other's autonomy. Thus, the principle of respect for autonomy, which at the birth of bioethics grew out of a specific concern to limit physician paternalism, becomes a general approach immuniz-ing patient choices from larger public scrutiny.[35] Whatever its virtues, however, such a theory is not rich enough to offer us much substantive guidance. It tends to mask our disagreements on crucial questions about suffering, human dignity, the mean-ing of death, and the relation of the generations. Beauchamp and Childress themselves do sometimes signal this fact to their readers, but it remains true that the architectonic of their system is deceptive in its clarity—leaving unaddressed the most pressing questions.

III

Perhaps the most philosophically developed alternative to the "applied ethics" approach of *Principles of Biomedical Ethics* is the revival of casuistry sponsored in particular by Albert Jonsen and Stephen Toulmin.[36] This approach has its roots, of course, in medieval Catholic moral theology—and perhaps, though I think the connections are more tenuous here, in Aristotle's notion of practical reason. As we shall see, how-ever, its more immediate roots lie in the concerns of twentieth-century Anglo-American moral philosophy.

A casuist does not begin his moral reflection at the level of theory, nor does he reason deductively. Rather, he begins from

cases about whose resolution most of us are relatively confident, cases that are "too clear and simple, too nearly paradigmatic to be in any way problematic or open to doubt."[37]
Reflection on a paradigm case will direct our attention to important moral features that may be at stake in some other,
seemingly more problematic, case. Reasoning by analogy, the
casuist seeks to decide whether we should treat the problematic case as we do the paradigm case. In this process of
analogical reasoning we might, in fact, appeal to a variety of
analogies or to different paradigm cases that pull us in several
directions. Gradually, by a process of weighing and reflecting
that certainly cannot be called deductive, we build a cumula-
tive case for treating the newer, problematic case in a certain
way. Thus, for example, if we are suddenly confronted with the
need to render moral judgment about surrogate motherhood,
we might initially consider how such an arrangement is best
described—in analogy with other contracts, as a form of baby
selling, as a kind of adoption, as just another form of artificial
insemination by donor, or as akin to circumstances in which
we consider the rights of a biological father who is not married
to the baby's mother. On at least some of these topics we may
already have some social consensus, and, thinking analogically,
we may manage to expand that consensus by applying it to the
new problem of surrogate motherhood. Such movement will
always be gradual and complex, of course, but the aim is a
modest one: to find an opinion that "could be reasonably
entertained."[38]

A willingness to proceed casuistically requires what both
Jonsen and Toulmin regard as a certain kind of moral maturity.
The popularity of the principlist approach to bioethics Jonsen
ascribes to a characteristically American "moralism" that cannot
live with complexity or ambiguity in the moral life, a moralism
that has religious roots in Calvinist thought.[39] Whether this is
meant as a historical account or a reading of the American

psyche is hard to say, but as the former it is doubtful. To be sure, Robert Veatch, himself a devoted advocate of a principlist approach to bioethics, has also traced the beginnings of bioethics to roots in Protestant theology. He suggests that "[i]mportant themes of rights and the affirmation of the lay decisionmaker read directly . . . from Protestant theological themes."[40] At least as plausible, however, is Daniel Callahan's claim that acceptance of the bioethics movement in this country depended in large part on its willingness to "push religion aside," divorcing bioethics from religion in the public mind. Rather than seeing the language of rights (and the related principlist approach) as growing out of Protestant theology, Callahan sees the turn to a language of rights as a move toward "the mainstream of public policy."[41] Moreover, that John Calvin "incorporated his entire moral theology under the rubric of the Decalogue" (a precursor, on Jonsen's account, of a principles approach to bioethics) hardly suffices to demonstrate that he is a "moralist" in Jonsen's sense.[42] After all, the most casual reading of Luther's catechisms would demonstrate that he made similar use of the Decalogue; yet few would describe him as one who, like Jonsen's moralist, denies "the possibility of moral paradox."[43]

In any case, on Jonsen's reading, the moralists gradually lost control of bioethics in this country. Especially in the work of the two national commissions, bioethics gradually fell into the hands of those who were able to appreciate "exceptions and consequences" more than "unbreachable rule and immutable value."[44] Such bioethicists were willing to settle for what the old casuists would have called "probabilistic" opinions that were not clearly wrong and might receive public support. Nonetheless, on Jonsen's account this triumph of the morally mature is not complete. For moralists remain among us, still in the grip of a "Calvinist anxiety" that makes them uneasy "in the presence of those more likely to allow exceptions than to uphold principles."[45] And a qualified remnant of this moralism remains in the continued attraction of principles in bioethics. "Autonomy, beneficence, non-maleficence and justice became the bioethicists' distant echoes of the Calvinist's Decalogue."[46]

This account is no more persuasive as psychology than as history; nonetheless, the renewal of casuistry raises serious issues about method in bioethics. If we consider the claims of Stephen Toulmin, Jonsen's fellow proponent of casuistry, we can see how these issues are rooted less in medieval casuistry than in twentieth-century Anglo-American metaethics. In 1950 Toulmin published *An Examination of the Place of Reason in Ethics*, which was reissued in 1986 under the slightly more assured title, *The Place of Reason in Ethics.*[47] The 1986 edition includes a preface situating his argument in relation to events after 1950 and, in particular, drawing connections with the birth of bioethics.

Toulmin hoped to overcome an impasse in moral theory between those who sought an objective foundation for ethics (usually in purported non-natural properties such as goodness or rightness) and those who held the more subjective view that moral judgments express our feelings or attitudes and do not, therefore, seek to state truths about the world. Toulmin sided with the subjectivists in holding that moral disagreements are not disagreements about facts, but he sought to support a central objectivist concern by arguing that moral disagreements are nonetheless genuine and are not just expressions of feelings. When we ask what ought to be done in some circumstances, he claimed, we are simply asking "whether there was any reason for choosing one course of action over another."[48] We need refer to no underlying principle or theory; we simply need to offer good reasons. Here, of course, is an anti-foundationalist position before the term became a popular one—even, it would not be too strong to say, an anti-theoretical approach. The attempt to offer "good reasons" ties moral reasoning closely to particular cases in which one attempts to discern both what is unique about the case and what may link it by analogy to other sorts of cases. In Toulmin's words, "the philosopher's task is not to find an underlying principle that binds all obligations and claims together; rather, it is to develop a sufficiently varied taxonomy of cases, circumstances, and considerations, allowing for (and doing justice to) the differences between them."[49] In short, the task is that of the casuist.

The appeal of such an approach will be greater, of course, if we are persuaded that the objectivist approach to moral reasoning is mistaken.[50] But in order to avoid relativism we may be drawn to objectivist positions—attracted to metaphors of "sight," which suggest that there are moral facts to be discerned just as much as other sorts of facts. Those who see them rightly discover truth; those who do not adopt false views of the moral life. Yet it may also be puzzling to say that "goodness" is a property of things, since people seem to disagree so much about where it is to be found. We tend to think that when things are there, they are there to be seen by everyone who looks. Could it be that some of us do not see well, that our organ of moral sight is deficient or distorted and that we suffer a kind of moral color blindness?

Something like that is the view of that great moral objectivist Plato, articulated in his myth of the cave.[51] The people in the cave are there from childhood, bound in such a way that they can look only in front of themselves. Behind them a fire burns, between them and the fire is a path along which statues of people, animals, and various objects are carried. Having been bound from birth, they see only the shadows cast by the fire from the objects carried along the path and hear only the sounds of the carriers. These shadows and echoes are, they think, reality. Some of them become quite adept at distinguishing among the shadows, and honors and prizes go to those skilled in this way. But deliverance from such distorted understanding comes only when a person is released and compelled to turn and walk upward toward the light. If we turn, we gradually come to see the realities whose shadows we had seen before, and we may even begin to look at the light. The turning is not complete, though, until we escape the cave and can look upon the sun.

Bound in the cave, we may become quite sophisticated in our ability to reason about cases in morality presented to us. But if the instrument through which we see has not been turned around toward goodness itself, this reasoning may be an exercise in futility. Some may impress us with their expertise and appear deserving of praise and prizes. But eventually it will become clear that our arguments cannot be settled and that, though the arguments may seem objective, they reflect not knowledge but our own preference or taste. If we approach moral reasoning by asking those in the cave what they see when they reflect upon cases, we forget, as Iris Murdoch has written, that "in opening our eyes we do not necessarily see what confronts us. We are anxiety-ridden animals. Our minds are continually active, fabricating an anxious, usually self-preoccupied, often falsifying *veil* which partially conceals the world."[52]

The image of knowledge here is of something that seeks almost to impose itself upon us, of "being transfixed by a sight which leaves us speechless."[53] Toulmin's image, by contrast, is not of sight but of conversation in which we offer our reasons, hoping others find them to be good reasons. Contemplating this contrast, we can understand why Toulmin rings the changes on a theme rather like Jonsen's: that a search for principles that bind universally is an anxiety-ridden quest for certainty in life.[54] Moral wisdom, he suggests, "is exercised not by those who stick by a single principle come what may, absolutely and without exception, but rather by those who understand that, in the long run, no principle—however absolute—can avoid running up against another equally absolute principle; and by those who have the experience and discrimination needed to balance conflicting considerations in the most humane way."[55] The function of ethics becomes, then, the achievement of social harmony. Appealing, as does Jonsen, to the experience of the National Commission as an example of the kind of moral reasoning we need, he argues for a revival of "Aristotelian procedures of the casuists and rabbinical scholars, who understood all along that in ethics, as in law, the best we can achieve in practice is for good-hearted, clear-headed people to triangulate their way across the complex terrain of moral life and problems."[56] In contrast to Murdoch's Platonic suggestion that we anxiety-ridden animals will not necessarily see the truth when we go looking for it, there is here a kind of complacent confidence that conversation between "good-hearted" people will not need any principles to protect them against the evils of the human heart.

Clearly, casuistry even more than principlism is a bioethical approach developed with public policy in mind. And when that is our aim, an absolute principle—a red light, and not just a yellow caution light—is always a barrier. Even if, therefore, theory is often idle in *Principles of Biomedical Ethics*, the aim of that volume to recognize the necessity of theory should be honored. Toulmin's early work in metaethics suggests that only a primitive ethic views the moral code in deontological fashion as fixed and unalterable. A more developed and urbane ethic, by contrast, thinks of a moral code as mandatory only in unambiguous cases. In all other cases, the development of the code is "controlled by appeal to the function of ethics; that is, to the general requirement that preventable suffering shall be avoided."[57] That there might be occasions when we could relieve suffering but ought not is a principle that will be hard to remember when we think in this way.

Indeed, in Jonsen's and Toulmin's revived casuistry it is not easy to determine where the unambiguous cases are to be found. They write of cases that are "too clear and simple, too nearly paradigmatic to be in any way problematic or open to doubt," but their examples are few.[58] The examples given at this point are cases of willful cruelty or taking cauliflower from the store without paying for it. At another point they speak of allowing no room for "conscientious disagreement about cases of, say, willful cruelty to innocents or purely selfish deceit."[59] When speaking of such paradigm cases as the beginning point for casuistic reasoning, Jonsen and Toulmin can, in fact, describe moral vision in "foundationalist" terms.

> In ethics as in mathematics we have an "eye" for paradigmatic or type cases: in both fields there is no appeal behind this experience. In unambiguous ("paradigmatic") cases we can recognize an action as, say, an act of cruelty or loyalty, as directly as we can recognize that a figure is triangular or square.[60]

Moreover, quoting words of Kenneth Kirk that they say "serve as a summary of the points about practical reasoning that we shall keep in mind," they note how casuistic reasoning tries to extend itself from a paradigmatic case to a newer, more puzzling case in such a way that, as the process continues, one might hope to reach "a definition so inclusive as to make further examination of instances superfluous. Then the law will be defined in relation to hitherto unforeseen areas in the map of conduct."[61] The reader may be a little surprised by this formulation, suggesting as it does that not only a few unambiguous paradigm cases but also more narrowly formulated judgments might be closed to future revision.

It would indeed be surprising were this the upshot of their method, and I do not think Jonsen and Toulmin can consistently commit themselves to such a description.[62] The truth is, rather, that even the paradigm cases are always subject to review. What we decide about newer cases may reverberate in ways that reshape what we once thought unambiguous.

Hence, "procedures for criticizing the core categories of moral practice are *built into* the established traditions of case reasoning. . . ."[63] Thus, for example, where we might once have thought that willful adultery was clearly wrong, we may be less certain if we now believe that the presence of love alone could render a sexual relationship permissible and praiseworthy. Believing that, we might have to revise our understanding of what had been paradigmatic. Thus, Jonsen and Toulmin write, "[n]othing at the heart of the so-called sexual revolution . . . weakens traditional moral objections to sexual relationships that are unloving and exploitive, or to promiscuity that is divorced from true human affection."[64] That formulation, even if accurate, must surely count as a radically revised version of what is paradigmatic and unambiguous. Of course, if we find on some issue—say, willful cruelty—"practical overlap" between what is taken as paradigmatic in many cultures, we may think reformulation unlikely, but even then "our certainty about the prescriptions of the core paradigms only differs in degree, not kind, from our confidence in the resolutions of more derivative cases."[65] For this view there can be, finally, no foundations—nothing all the way down except good-hearted people triangulating their way through moral complexity.

Clearly a *theoretical* decision has been made, a commitment to consequentialism undertaken, even if the theory is very unlike classical utilitarianism. And, beyond any doubt, those who formulate public policy must in large measure be governed by results. If that is to be the task not just of our legislators but of bioethicists as well, if we are to have a bioethics framed with one eye (or perhaps both) on the requirements of public policy, there is a good bit to be said for principlism rather than casuistry. For the principlist approach is at least open to deontological considerations. Within its framework one can find ways to say "go no farther," not just to say "proceed with caution." Against the charge that principlism has developed a

"regulatory bioethics for the society rather than offering prophetic judgments," James Childress has noted that principlists
have been able to set themselves against some features of our
health care setting—against, for example, professional paternalism.[66] And he suggests, quite reasonably I think, that principles
can "serve a critical function, perhaps more readily than the tax-
onomic approach of a pure casuistry that attempts to operate
without principles."[67] The question that remains, however, is
whether bioethics might better turn in another direction, no
longer seeing its task chiefly in terms of shaping public policy.

IV

That casuistry has seemed to offer a usable method in bioethics
(as, for example, Jonsen claims it did within the National Commission) may be due largely to the fact that medical practice
is something of an anachronism within our culture. It has
retained a greater coherence and integrity than many of our
institutions, and, hence, it still retains some acknowledged
wisdom about how to act in standard situations. Because its coherence may be somewhat greater than that of the surrounding
culture, one might cling to the hope that casuistry could offer a
serviceable method. But it is much more likely that the frag-
mentation of our culture will continue to erode the integrity of
medical practice, making casuistry increasingly difficult. Jonsen
and Toulmin themselves recognize that their method "requires
institutions that provide the locus for and lend support to the
uniquely casuistical way of approaching moral problems."[68] In
the absence of cultural circumstances that might actually shape
and sustain such institutions, however, we must simply argue
over the fundamental rules of life and the basic purposes of a
social practice such as medicine.[69]

This would mean taking as our principal aim not the achievement of social harmony on bioethical questions but the probing of issues in ways that give expression to our understandings of human nature and moral excellence. If we did not take consensus as our aim, we would less easily be driven to make autonomy central in bioethics. That move inevitably turns bioethics in a procedural rather than substantive direction; for, lacking substantive agreement there remains only the search for ways to honor everyone's autonomy. With a different aim we would be less likely to confuse the ideal of the autonomous person (about which we should be skeptical) with the principle of respect for autonomy (which deserves support when given nuance within a fuller vision of the human person). Bioethics might then "expand its own horizons" by returning to the metaphysical richness that characterized its early years.[70]

In this undertaking we need not eschew principles, though our method should perhaps be neither the "application" of principles to cases nor the intuitive "balancing" of conflicting principles, but, instead, what Henry Richardson has termed "specification."[71] The defects of application and balancing are well known. Suppose, for example, that we are considering whether in certain circumstances a doctor ought to deceive or even lie to a patient. The application model requires that we systematize the several principles involved in a way that precludes conflict among them, since without such systematization one cannot know what principle to apply. This is, at least many have supposed, "an ideal achievement of moral theory that is beyond our grasp."[72] If application seems unattainable, intuitive balancing inevitably seems arbitrary and without rational grounding.

Richardson's concept of specification, by contrast, asks us to proceed by "qualitatively tailoring our norms to cases."[73] When our normative commitments appear to conflict, we do not simply change our mind, but we qualify our commitments in such a way that at least one of them is made more specific.

What Richardson has in mind is not unlike a suggestion made by Paul Ramsey in a long essay treating the problem of exceptional cases in moral reasoning.[74] Ramsey took up for discussion a case used by Joseph Fletcher to illustrate the need for situation ethics—the case of Mrs. Bergmeier's "sacrificial adultery." Confined by the Russians in a concentration camp after the collapse of Nazi Germany, Mrs. Bergmeier was desperately needed by her husband and three children in those difficult and dangerous times. Camp regulations provided that a woman inmate who was pregnant would be returned to Germany, since she was considered a liability in the camp. Therefore, seeing no other way of return, Mrs. Bergmeier asked a friendly guard to impregnate her. She was released to rejoin her family, who were overjoyed to receive her and the newest member of their family.

Ramsey considers several approaches we might take to justify Mrs. Bergmeier's action. The one that interests us here is his suggestion that such a case "enables us to make explicit certain meanings and stipulations that were all along *implicit* in the meaning of *marital fidelity*."[75] That is, as we ponder the case, we do not so much understand Mrs. Bergmeier's action as exempted from the requirement of marital fidelity as we begin to deepen and clarify our own concept of such fidelity. Mrs. Bergmeier need not have appealed to some other principle, nor need she have appealed to what seemed best on the whole. She need only have considered the requirement of marital fidelity and then have "extended into practice in a critical situation what *this* entailed. . . . The extraordinary situation she faced brought to light certain stipulations or understandings that were implicit in the requirement of this specific fidelity."[76] Her action was not therefore adulterous but was justified within the meaning of marital fidelity.

This is only one of the avenues for moral reasoning that Ramsey contemplates in such circumstances, and he is not fully persuaded that it will work in the case of Mrs. Bergmeier, but it gives us a sense of what it might mean to attempt such specification that qualitatively tailors norms in the light of particular cases. Ramsey's work in bioethics often used such a method. Thus, for example, in the classic chapter "On (Only) Caring for the Dying," in *The Patient as Person*, Ramsey argued that it was our duty never to abandon care of the dying (which did not of course mean that treatment could not sometimes be withdrawn precisely in the name of care).[77] At the end of the chapter, however, Ramsey proposed for consideration two "qualifications" to our duty never to abandon care.[78] Suppose there were (about which possibility Ramsey remained agnostic) patients who were irretrievably inaccessible to care or whose pain could not be relieved and was so intense as to make them utterly beyond the reach of our care. At such a point, his suggestion was, hastening death need no longer be understood as a violation of our duty. For if we specify adequately what it means to be indefectible in our care, we will see that what cannot be received is not part of the meaning of "care." Later in his career Ramsey largely withdrew these exceptions to the duty never to abandon care, but the example of his procedure—exploring and deepening the meaning of "care"— remains a useful one.[79]

Such exploration invites precisely the richer and deeper reflection upon the meaning of human life that a bioethics aimed at lowest common denominator public policy does not provide. Thus, for example, when Ramsey took up the issue of organ donation in order to ponder possible justifications of the self-giving of vital organs, he had to face the fact that the language of Christian love might appear to endorse such heroic self-giving.[80] He had therefore to consider how "to conceive of charity as the action of creaturely men, of men of flesh, and not as the action of disembodied spirits."[81] He had to reflect

upon the place of the body in our understanding of personhood. He had to consider whether human beings "can manifest the fact that there are limits to their *refusal* to accept the
death of a dear one in the flesh."[82] He had, that is, to reflect
upon the meaning of suffering, aging, and dying.

One of the advantages of such an approach is that it will
invite religion back into bioethical inquiry. Not all will welcome
such a development, of course. Discussing the birth of bioethics
at which she was present, Shana Alexander writes that the study
of ethics at that time had not really progressed from "ancient
times."[83] One hardly knows what to make of such a statement,
of course, but she means, evidently, that ethics was still "im-
mersed in religion." In her view, bioethics made progress because the rapid growth of technology brought into the field
others who were willing to move beyond religious positions.

This much is true: Many of the early figures in the bioethics
movement were scholars in the field of religion, and in the several intervening decades bioethics has largely fallen into the
hands of scholars trained in other disciplines. Robert Morison
once wrote that as late as 1981 the developing field of bioethics
was still more dominated by theologians than by secular philosophers.[84] I doubt if such dominance still existed even then;
certainly it does no longer. Indeed, Daniel Callahan has suggested that bioethics gained public acceptance by pushing
religion aside (even if unintentionally).[85] In its place he detects
movement toward "a different kind of moral language in the
mainstream of public policy, toward a language of rights" that
seeks "moral consensus . . . in the face of a diverse cultural situation."[86] That movement we have traced to some degree in this
chapter and have concluded, as Callahan himself does, that
bioethics may need to return to its earlier self, expanding its
horizons and no longer understanding its function chiefly in
terms of social consensus.

This will mean, in part, inviting back those "alternative
imaginations" that religious communities and theological tra-

ditions provide.[87] The same principles that Beauchamp and Childress put forward may still often shape our discussion, but they will take on new resonance. The selves whose autonomy we respect will be understood as grounded in community   and in relation to God. The imperatives of beneficence may sometimes seem too minimal. Our sense of justice will be constantly reshaped by concern for those who are weak and cannot speak in their own behalf.[88] Of course, such approaches may sometimes ask more of our fellow citizens than is possible, or more than we ourselves can always be persuaded to undertake, but at least we will not have begun our reflection with the intent of seeking no more than public policy currently envisions. If compromise and adjustment are necessary in our common life, that can and should be left to the processes of democratic governance—a politics that, because it does not claim our souls, paradoxically can allow matters of the soul into public argument.

As a way of concluding these reflections on method in bioethics, I turn finally to Leon Kass's powerful critique of the present practice of bioethics. In "Practicing Ethics: Where's the Action?" a talk given at the celebration of the Hastings Center's twentieth anniversary, Kass drew up what can only be called a powerful indictment of the bioethics movement.[89] Through considering his central concerns, I will try to clarify my own claims.

Kass is critical, first, of the move toward principles and rules in bioethics. "No guidelines can cover all real cases, much less touch the critical nuances that distinguish any one case from another. The methodical rationality of procedure is put in place of the discerning reasonableness of the prudent man-on-the-spot that all real choices demand."[90] Kass himself, of course, defends the view that there are virtues intrinsic to the practice of medicine that are not simply the application to medicine of a more general moral standpoint. I have granted above that this is true in part, but only in part. The moral vision internal to the practice of medicine must always be in conversation with our more general principles, even if there is sometimes considerable tension between them. Moreover, while it is true that the application of principles to cases always requires the insight of prudential wisdom, it is equally true that we cannot reason morally without beginning to generalize about the characteristics that connect cases to each other. And at certain points, at least, we cannot avoid theoretical argument. With respect to the ethic internal to medicine and with respect to the tensions between that ethic and the larger society's concerns, we will not be able to avoid the conflict between consequentialist and deontological theories. We can, to take an example, grant that attacks made in the name of patient autonomy on medical paternalism may have underrated the importance of the "discerning reasonableness" of physicians without conceding that bioethics can be divorced from larger questions of moral theory. To the degree that Kass's first criticism sides with the casuists against the principlists, it does not yet get to the core of our problems. There is still a place for principles and theory in bioethics.

Kass argues, second, that greater theoretical precision has not necessarily improved the practice of bioethics. "Bioethicists have by and large behaved as if they could (and should) do no more than give pious blessings to the inevitable."[91] Certainly this is not an adequate description of everything that has happened within the bioethics movement. If Rothman is right that the movement began as a kind of offshoot of the larger civil rights movement, if we recall, for example, the continuing critique of abuses in human experimentation, we can see that bioethicists have sometimes set themselves against practices widely accepted. Nevertheless, as bioethics became oriented more toward the shaping of public policy (through the several national commissions), it has in some respects been domesticated. If Kass's first concern tilted him in the direction of the casuists, this second concern certainly does not. For, as we have seen, Jonsen and Toulmin are quite clear that their method aims at consensus and is exemplified by reports of the several national commissions. If those reports do not simply "give pious blessings to the inevitable," it still would not be wrong to describe them as somewhat domesticated. And although Kass sets himself strongly against the principlist approach, on this point I would note, as I did earlier, that principles may serve a critical function—perhaps more readily than a taxonomy of cases.

Kass's third criticism of the present state of bioethics returns us to the claims of Ezekiel Emanuel. Kass argues that our understanding of the relation of theory and practice is inadequate, that the metaphor of "applying" theory to practice—or principles to cases—does not go deep enough to uncover our true difficulties. Thought can have an effect only if it is somehow united with appetite. Without appetites already directed toward what is good, we will not be in a position usefully to discuss theories or principles. This leads Kass to suggest that we must begin with practice, and that, in turn, in good Aristotelian fashion, leads him from ethics to politics. "It may turn out that changes in divorce law or childcare practices are ethically far more deserving of our attention than arguments about the status of the *in vitro* embryo or the rights of its biological progenitors. It may turn out that designing programs of compensated national service for our high school graduates deserves as much of our ethical attention as the ethics of various techniques of behavior modification."[92] This is actually a more faithful use of Aristotle than Emanuel's attempt to focus on groups that dealt with questions of medical practice alone. And Kass rightly understands that, to the degree Aristotle is correct in holding that desire must become thoughtful or intellect must be suffused with appetite, moral education is a crucial prerequisite for good moral reasoning about bioethics. I, at least, believe that Kass—not to mention, Aristotle!—is essentially right on this point, but we have to ask what it means for bioethics. It might challenge each of us to consider whether our efforts could better be turned in other directions. But the fact that childcare practices are more important than arguments about the status of the *in vitro* embryo—a point that I grant even while being uncertain just how one makes such determinations—does not mean that the status of the embryo is unimportant. Nor does it mean that all bioethicists should shift to another vocation. Nor, precisely because ethics and politics are not the same, must bioethics give way entirely to larger political questions. It has—or should seek to have—its own integrity, even if, as must surely be true, its health cannot be separated entirely from the moral health of our community.

Finally, Kass notes that the people who founded the Hastings Center and the bioethics movement more generally did not come to it "through either the study or the practice of bioethics as it is now practiced and studied."[93] We ought to be worrying, he suggests, about how we will get more of the sorts of persons with the sorts of concerns that first energized the movement, and standard training in bioethics may not accomplish this. "We must return to what animated the enterprise: the fears, the hopes, the repugnances, the moral concern, and, above all, the recognition that beneath the distinctive issues of bioethics lie the deepest matters of our humanity."[94] Here, I think, is the heart of the matter. Questions of method are important at certain points, and, indeed, they cannot entirely be avoided. But what finally counts is whether a method opens bioethical reflection to these "deepest matters of our humanity." For, in fact, what is often at stake in bioethics is precisely that humanity. No method can ignore that and retain the soul of bioethics—or our own soul.

2. How Bioethics Lost the Body: Personhood

A bioethics that, in its attempt to forge public policy, has lost the soul—that has left behind the alternative imaginations provided by religious vision—has, in a more literal sense, lost the body. In this chapter and the next I explore two different ways in which bioethical reflection has begun to lose the moral significance of our bodies. Here I will focus on some of the issues that have emerged in discussions of death, dying, and care for the dying—as a way of thinking about what it means to have a life.

In particular, I will focus on a concept that has risen to great prominence in our thinking—the concept of a person. Two competing visions of the person—and the relation of person to body—have unfolded as bioethics has developed, and, in my view, the wrong one has begun to triumph. We have tried to handle our substantive disagreements on this question by turning to procedural solutions—in particular, advance directives—trusting that they presume no answer to the disputed question. We are, however, beginning to see how problematic such a procedural solution is—how flawed and, even, contradictory much thinking about advance directives has been. What we need, I will suggest, is to recapture the connection between our person and the natural trajectory of bodily life.

That will be the course of my argument. But, as a way of framing the issues, I begin in what is likely to seem a strange place: with the thought of some of the early Christian Fathers about heaven and the resurrection of the dead. They were attempting to relate the body's history to their concept of the person's optimal development. In so doing, they provide a different and illuminating angle from which to see our present concerns.

I

In his *City of God* Saint Augustine describes the human being as *terra animata*, "animated earth."[1] Such a description, contrary in many ways to trends in bioethics over the last several decades, ought to give pause to anyone inclined to characterize Augustine's thought simply in terms of a Neoplatonic dualism that ignores the personal significance of the body. It may, in fact, be our own constant talk of "personhood" that betrays a more powerful tendency toward dualism of body and self.

This same Augustine, however, found himself puzzled at the thought of the resurrected body. What sort of body will one who dies in childhood have in the resurrection? "As for little children," Augustine wrote, "I can only say that they will not rise again with the tiny bodies they had when they died. By a marvellous and instantaneous act of God they will gain that maturity they would have attained by the slow lapse of time" (22.14). This is, in fact, a question to which a number of the Church Fathers devoted thought.[2]

Origen, for example, understood that throughout life our material bodies are constantly changing. How, then, can the body be raised? He appealed (in good Platonic fashion) to the *eidos*, the unchanging form of the body. It remains the same as we grow from infancy, through childhood and adulthood, to old age. Hence, despite the body's material transformations, its *eidos* remains the same throughout. (For Origen this *eidos* is not the soul; it is the bodily form united with the soul in this life and, again, in the resurrection. J.N.D. Kelly comments that Origen was charged with having held that resurrected bodies would be spherical—and may have held such a view, in keeping with the Platonic theory that a sphere is the perfect shape.)

From here it is not a long step to suppose that, since the *eidos* of each resurrected body will be perfect, it will in every instance be identical in qualities and characteristics. Thus, Gregory of Nyssa, though differing from Origen in some respects, held that in the resurrection our bodies will be freed from all the consequences of sin—including not only death and infirmity, but also deformity and difference of age. This is a view not unlike Augustine's. Bodies may have a (natural) history, but the bodily form is unchanging. That form is the human being at his or her optimal stage of development, the person as he or she is truly meant to be. (I write "he or she" not simply to conform to current canons but because Augustine, for example, took trouble to note that the sexual distinction—but not the lust that, in our experience, accompanies it—would remain in the resurrection. All defects would be removed from the resurrected body, but "a woman's sex is not a defect" (*CD* 22.17). And although intercourse and childbirth will be no more in the resurrection, "the female organs . . . will be part of a new beauty." This is perhaps what C. S. Lewis had in mind when he wrote: "What is no longer [in the resurrection] needed for biological purposes may be expected to survive for splendour."[3]

Against Origen's notion that the resurrected body would be a purely spiritual *eidos*, Methodius of Olympus held that the body itself—not just its form—would be restored in the resurrection. He based his claim less on a developed philosophical argument than on the Resurrection of Jesus, who was raised in the same body that had been crucified (complete, we may recall, with the nail prints in his hands). Such issues continued to occupy the attention of theologians for centuries to come. Thus, Saint Thomas considers the nature of the resurrected body. For Thomas, the form of the body is the rational soul, and the body reunited with that soul in the resurrection need not reassume all the matter that had ever been its own during temporal life. Rather, Thomas suggests in the *Summa Contra Gentiles*, the resurrected man "need assume from that matter only what suffices to complete the quantity due. . . ."[4] The "quantity due" is whatever is "consistent with the form and species of humanity." This means that if one had died at an early age "before nature could bring him to the quantity due," or if one had suffered mutilation, "the divine power will supply this from another source" (4.81.12). Saint Thomas is emphatic—against what may have been Origen's view—that our risen bodies will not be purely spiritual. Like Christ's they will have flesh and bones, but in these bodies there will not be "any corruption, any deformity, any deficiency" (4.86.4). Nor, it appears, will there be differences of age; for all will rise "in the age of Christ, which is that of youth [young adulthood], by reason of the perfection of nature which is found in that age alone. For the age of boyhood has not yet achieved the perfection of nature through increase; and by decrease old age has already withdrawn from that perfection" (4.88.5).

At least to my knowledge, this sort of speculation becomes much rarer after the Reformation—perhaps because Protestants were less inclined to go beyond biblical warrants, even when an intriguing and potentially significant question beckoned. In the fifteenth and last of his charity sermons, Jonathan Edwards does say of heaven: "There shall be none appearing with any defects either natural or moral."[5] And more recently Austin Farrer has approached these questions by asking how it is possible for us to "relate to the mercy of God beings who never enjoy a glimmer of reason."[6] If there never was a speaking and loving person, Farrer asks, where is the creature for God to immortalize? He is less troubled by those who have lost the speaking and loving personhood that once was theirs; God can immortalize them, though Farrer does not tell us whether they are immortalized free of defects or, even, age differences. But what of those in whom reason never developed? "The baby smiled before it died. Will God bestow immortality on a smile?" Farrer contemplates, without being satisfied by, the possibility that "every human birth, however imperfect, is the germ of a personality, and that God will give it an eternal future"—a speculation not entirely unlike that of some of the early Fathers. And he realizes that there may be some who, though retarded, are not completely without reason—though he never asks, then, what sort of eternal future might be theirs.

If we can overcome both our Enlightened bemusement at such speculation and our Protestant urge toward simplicity that refuses to learn from questions which admit of no answer, if instead we enter into the spirit of such questioning, we may find ourselves rather puzzled. Could such a monochromatic heaven—all of us thirty-five years old, well endowed with (identical?) reasoning capacities—really be heavenly? If each of the saints is to see God and to praise the vision of God that is uniquely his or hers, and if the joy of heaven is not only to see God but to be enriched by each other's vision, then why should we not look through the eyes of persons who are very different indeed? Is not the praise of a five-year-old different from that of a thirty-five-year-old, and, again, from that of a seventy-five-year-old? Why should not these distinct and different visions be part of the vast friendship that is heaven? Perhaps it is easier to understand the tendency to eliminate any defects from heaven, but even there—when they closely touch personal identity—we may find ourselves rather puzzled. Edwards was, for example, confident that there would be neither moral nor *natural* defect in heaven. Yet, he was willing to grant that friends will know each other there. But if the stump that should have been my leg has shaped the person who I am, the person who has been your friend for forty years, it is hard to know exactly what our heavenly reunion is to be like when the stump is replaced by a perfectly formed leg. "Will God bestow immortality on a smile?" As likely, I should think, as that the mother of that child will meet one upon whom God has, in Augustine's words, bestowed in "a marvellous and instantaneous act . . . that maturity they would have attained by the slow lapse of time." We might set against Farrer's view the comment of his fellow Anglican David Smith, who writes that "at the very least it would be hard for Anglicans to hold that a being who might be baptized was lacking in human dignity."[7]

Perhaps I begin to wax too enthusiastic in my own speculations, but the point is worth pondering. To live the risen life with God is, presumably, to be what we are meant to be. It is the fulfillment and completion of one's personal history. To try to think from that vantage point, therefore, is to imagine human life in its full dignity. And to try, however clumsy the speculation, to adopt this vantage point for a moment is to think about what it means to have a life. The questions I have been considering invite us to think about our person, our individual self. Does it have a kind of timeless form? A moment in life to which all prior development leads and from which all future development is decline? A moment, then, in which we are uniquely ourselves? Or is our person simply our personal history, whether long or short, a history inseparable from the growth, development, and decline of our body?

There is some reason to think—or so I shall suggest in what follows—that much contemporary thought in ethics has a great deal in common with Origen's thinking. In an age supposedly dominated by modes of thought more natural and historical than metaphysical, we have allowed ourselves to think of per-

sonhood in terms quite divorced from our biological nature
or the history of our embodied self. In the words of Holmes
Rolston, our "humanistic disdain for the organic sector" is
"less rational, more anthropocentric, not really *bio*-ethical at
all," when compared to a view that takes nature and history
into our understanding of the person.[8] Or, put in a more liter-
ary vein, the view I will try to explicate is that expressed by Ozy
Froats in Robertson Davies' novel *The Rebel Angels*. Froats, a
scientist, is discussing his theories about body types with Simon
Darcourt, priest and scholar. Froats believes there is little one
can do to alter one's body type, a dismaying verdict for Darcourt, who had hoped by diet and exercise to alter his tendency
toward a round, fat body. Froats says of such hopes.

> To some extent. Not without more trouble than it would
> probably be worth. That's what's wrong with all these diets
> and body-building courses and so forth. You can go against
> your type, and probably achieve a good deal as long as you
> keep at it. . . . You can keep in good shape for what you are,
> but radical change is impossible. Health isn't making every-
> body into a Greek ideal; it's living out the destiny of the body.[9]

Terra es animata.

II

Ozy Froats' notion of having a life is not, however, the vision
that seems to be triumphing in bioethics. And, to the degree
that developments in bioethics both reflect and shape larger
currents of thought in our society, those developments merit
our attention.

The language of personhood has been central to much of
the last quarter century's developments in bioethics. It was
there at the outset, when in 1972, in the second volume of *The*

Hastings Center Report, Joseph Fletcher published his "Indicators of Humanhood: A Tentative Profile of Man."[10] The
language had not yet solidified, since Fletcher could still use
human and *person* interchangeably. But the heart of his view
was precisely that which would, in years to come, distinguish
clearly between the class of human beings and the (narrower)
class of persons.

Among the important indicators (by 1974 Fletcher would
declare it fundamental)[11] was "neo-cortical function." Apart
from cortical functioning, "the *person* is non-existent." Having
a life requires such function, for "to be dead 'humanly' speaking is to be ex-cerebral, no matter how long the body remains
alive." And, in fact, being a person has more to do with being
in control than with being embodied. Among the indicators
Fletcher discusses are self-awareness, self-control (lacking which
one has a life "about on a par with a paramecium"), and control
of existence ("to the degree that a man lacks control he is not
responsible, and to be irresponsible is to be subpersonal").
Human beings are neither essentially sexual nor parental, but
the technological impulse *is* central to their being ("A baby
made artifically, by deliberate and careful contrivance, would be
more *human* than one resulting from sexual roulette. . . .").

Even if, in the briskness with which he can set forth his
claims, Fletcher makes an easy target, he was not without considerable influence—and it may be that he discerned and
articulated where bioethics was heading well before the more
fainthearted were prepared to develop the full consequences of
their views. Certainly the understanding of personhood that
he represents is very different from Augustine's "animated
earth" or Ozy Froats' sense that one must live out the destiny
of the body. Views of that sort have come to be labeled "vitalism," and their inadequacy assumed.

This is especially evident in our attitude toward death and
toward those who are dying. To confront our own mortality

or that of those whom we love is to be compelled to think about our embodiment and about what it means to have a life.[12] How we face death, and how we care for the dying, are not just isolated problems about which decisions must be made. These are also occasions in which we come to terms with who we are, recognizing that we may soon be no more. The approach of death may seem to mock our pretensions to autonomy; at the least, we are invited to wonder whether wisdom really consists in one last effort to assert that autonomy by taking control of the timing of our death. Contemplation of mortality reminds us that our identity has been secured through bodily ties—in nature, with those from whom we are descended; in history, with those whose lives have intertwined with ours. We are forced to ask whether the loss of these ties must necessarily mean the end of the person we are. Such issues, fundamental in most people's lives, have been involved in arguments about how properly to care for the dying, as we can see if we attempt to bring to the surface two contrasting views within bioethics about what it means to have a life.

For some time the distinction between "ordinary" and "extraordinary" care dominated bioethical discussions of care for the dying. It provided categories by which to think about end-of-life decisions. When this language began to be widely used—and, indeed, it did filter quite often into ordinary, everyday conversation—its chief purpose was a simple one. The perception—in many ways accurate—was that patients needed moral language capable of asserting their independence over against the medical establishment. They needed to be able to have ways of justifying treatment refusals, ways of resisting overly zealous—even if genuinely concerned—medical caregivers. A widespread sense that patients found themselves confronting a runaway medical establishment lay behind arguments that "extraordinary" or "heroic" care could rightly be refused and that no one had a moral obligation to accept such care. Over against a runaway and powerful medical establishment, this language sought to restore a sense of limits and an acceptance of life's natural trajectory. The language proved inadequate, however, meaning too many different things to different people. But it was not simply inadequate; it was also a language that did not, taken by itself, accentuate the increasingly prominent concept of personhood. And that concept has been used to broaden significantly the meaning of "useless" or "futile" treatment, by divorcing the person from the life of the body.

In recent years we have seen a spate of articles seeking to define futility in medical care. Care that is futile or useless has in the past been considered "extraordinary"—and could be refused or withheld. But what do we mean by futility? Years ago, when I was younger and more carefree, I used to enjoy going out at night in the midst of a hard snowstorm to shovel my driveway. In a sense, this was far from futile, since its psychological benefits were, I thought, considerable. But if the aim was a driveway clear of snow, it was close to futile. Well before I had finished, if the snow was coming hard, the driveway would again be covered. And sometimes I'd do it again before coming in—though aware that those inside were laughing at me. But if the goal was a driveway clear of snow, it just could not be accomplished, no matter how hard I worked while the snow was falling. "In Greek mythology, the daughters of Danaus were condemned in Hades to draw water in leaky sieves. . . . A futile action is one that cannot achieve the goals of the action, no matter how often repeated."[13]

This sense of futility we all understand, even if we realize that it may be difficult to apply with precision in some circumstances. Thus, for example, the comatose person (unlike the person in a persistent vegetative state [PVS]) is reasonably described as "terminally ill." Because the cough, gag, and swallowing reflexes of the comatose patient are impaired, he or she is highly susceptible to respiratory infections and has a life span usually "limited to weeks or months."[14] Because these reflexes are not similarly impaired in the PVS patient, he or she may live years if nourished and cared for. It makes sense, therefore, to describe most medical care for the comatose person as futile, and we understand readily, I think, the language of futility in that context. It is not as obvious, however, that the same language is appropriate in referring to the PVS patient.

Recent discussions make clear that, in light of such problems, "futility" has gradually come to mean something else—and something quite different. If the sense of futility described above is termed "quantitative" (referring to the improbability that treatment could preserve life for long), a rather different sense of futility is now termed "qualitative." Thus, some have argued, treatment that preserves "continued biologic life without conscious autonomy" is qualitatively futile.[15] It is effective in keeping the earth that is the body animated—effective, but, so the argument goes, not beneficial, because what is central to being a person cannot be restored.

How schizophrenic we remain on these questions becomes evident, however, when we contrast that view with a recent article on "New Directions in Nursing Home Care."[16] The authors argue that the standard view of autonomy that has governed so much of our thinking about acute care in the hospital context is not applicable to the nursing home patient. There we need a new notion of "autonomy within community." This may not be the best language to make their point, however, since the authors want to do more than just envision the person within his community of care. They are also concerned to see his medical condition, his chronic needs, his dependence, as internal to the person. Thus, they seek a "notion of moral personhood that is not abstracted from the individual's social context or state or physical and mental capacity. . . . For now the caring constitutes the fabric of the person's life . . . and the reality of the moral situation is that the person must embrace dependency rather than resisting it as a temporary, external threat."

The aim here is no longer returning the patient to an autonomous condition, having fended off the threat external to his person; instead, the aim is to rethink autonomy, to take into it a loss of self-mastery, to accept dependence in order "to give richer meaning to the lives of individuals who can no longer be self-reliant." Perhaps we might even say that the aim is to help the chronically ill person live out the destiny of the body.

How can it be, in essentially the same time and place where this argument is put forward, that we should be moving rapidly away from such an understanding of the person in so many discussions of "futile" medical care? When Dr. Timothy Quill assisted his patient, Diane, to commit suicide, he did it, he said, in order to help her "maintain . . . control on her own terms until death." The hands are the hands of Dr. Quill, but the voice is that of Joseph Fletcher, an increasingly powerful voice in our society.

Around the time that Fletcher was publishing his indicators of humanhood, one of the other great figures in the early years of the bioethics movement, also a theologian, was writing that the human being is "a sacredness in the natural biological order. He is a person who within the ambience of the flesh claims our care. He is an embodied soul or an ensouled body."[17] In those words of Paul Ramsey, the vision of the human being as *terra animata* was forcefully articulated.

As "embodied souls" we long for a fulfillment never fully given in human history, for the union with God that is qualitatively different from this life—which longing can never, therefore, be satisfied by a greater quantity of this life. But as "ensouled bodies" our lives also have a shape, a trajectory, that is the body's. Our identity is marked, first, by the bodily union of our parents, a relationship that then gradually takes on a history. We are a "someone who"—a someone who has a history—and though we may long for that qualitatively different fulfillment, we never fully transcend the body's history in this life. To come to know who we are, therefore, one must enter that history.

It is a history that may be cut short at any time by accident or illness but which, in its natural pattern, moves through youth and adulthood toward old age and, finally, decline and death. That is the body's destiny. As Hans Jonas has suggested, we exist as living bodies, as organisms, not simply by perduring but by a constant encounter with the possibility of death.[18] We constantly give up the component parts of our self in order to renew them, and our continued life always carries within itself the possibility that these exchanges may fail us. Eventually we are worn down, unable any longer to manage the necessary exchanges. The fire goes out, and we are no longer "animated" earth.

To point to some moment in this history as the moment in which we are most truly ourselves, the vantage point from which the rest of our life is to be judged—a moment at which, presumably, we have personhood, and not just another of the many moments in which we are persons—is to suppose that we can somehow extricate ourselves from the body's natural history, can see ourselves whole. It is even, perhaps, to suppose that in such a moment we are rather like God, no longer having our personal presence in the body.

It is not too much to say that two quite different visions of the person—Fletcher's and Ramsey's—have been at war with each other during the three decades or so that bioethics has been a burgeoning movement. But it is equally clear that one view has begun to predominate within the bioethics world and perhaps within our culture more generally. Among the peculiarities of our historicist and purportedly anti-essentialist age is the rise to prominence of an ahistorical and essentialist concept of the person. On this view, it is not the natural history of the embodied self, but the presence or absence of certain capacities, that makes the person. Indeed, we tend to think and speak not of being a person but of having personhood, which becomes a quality added to being. The view gaining ascendancy does not think of dependence or illness as something to be taken into the fabric of the person and lived out as part of one's personal history. It pictures the real person—like Origen's spherical *eidos*—as separate from that history, free to accept or reject it as part of one's person and life. Moreover, to be without the capacity to make such a decision is to fall short of personhood.

This view is not required by any of the standard approaches to bioethical reasoning or any of the basic principles (such as autonomy, beneficence, and justice) so commonly in use. What we do with such principles depends on the background beliefs we bring to them. Those beliefs determine how wide will be the circle of our beneficence, and whether our notion of autonomy will be able to embrace dependence. The problems we face lie less with the principles than with ourselves. We have lost touch with the natural history of bodily life—a strange upshot for *bio*ethics, as Holmes Rolston noted. How wrong we would be to suppose that ours is a materialistic age, when everything we hold central to our person is separated from the animated earth that is the body.

III

It might be, however, that I have overlooked something important in the paragraphs above. If in some cases we judge care futile when the capacity for independence is gone, and if in other cases of chronic illness we take the need for continual care into the very meaning of personal life, perhaps—one might suggest—the difference lies in what different people want, how they choose to live. One patient chooses to live on; another sees no point in doing so. Hence, the key is autonomous choice, which remains at the heart of personhood. All we need to do is get people to state their wishes—enact advance directives—while they are able. Then, if the day comes when others must make decisions for them, we will not have to delve into disputed background beliefs about the meaning of personhood. We will have a procedure in place to deal with such circumstances.

In the wider sweep of history, living wills are a very recent innovation, but the debate about their usefulness or wisdom coincides with the quarter century in which bioethics has grown as a movement.[19] And when we are told that, within a month after the Supreme Court's *Cruzan* decision, one hundred thousand people contacted the Society for the Right to Die, seeking information about living wills, we can understand that this is not an issue for specialized academic disciplines alone. The term *living will* was coined in 1969, and the nation's first living will law (in California) was passed in 1976—prompted, it seems, by the Karen Quinlan case. By now most states have enacted laws giving legal standing to living wills, and in 1991 the federal Patient Self-Determination Act went into effect, requiring hospitals to advise patients upon admission of their right to enact an advance directive. In a relatively short period of time, therefore, the idea of living wills (and other forms of advance directives, such as the health care power of attorney) seems to have scored an impressive triumph. If we have no substantive agreement on what it means to be a person or have a life, the living will offers a process whereby we can deal with such substantive disagreement. Each of us autonomously decides when our life would be so lacking in personal dignity as to be no longer worth preserving—and we pretend that such a process masks no substantive vision of what personhood means.

But it does, of course. Such a procedural approach brings with it a certain vision of the person: To be a person is to be, or have the capacity to be, an autonomous chooser, to take control over one's personal history, determining its bounds and limits. This substantive view turns out to have a life of its own and—we are beginning to see—can lead in several quite different directions. For a time, perhaps, all choices of once autonomous patients are honored. You choose to die when your ability to live independently and with "dignity" wanes; I choose to live on even when my rational capacities are gone. Each of us is treated as he has stipulated in advance. But then a day comes—and, indeed, is upon us—when the vision of the person hidden in this process comes to the fore.

If the person is essentially an autonomous chooser, then we will not forever be allowed to choose to live on when our personhood (so defined) has been lost. Living wills had, for the most part, been understood as a means by which we could ensure that we were not given care we would no longer have wanted, care that preserved a life regarded as subpersonal and no longer worth having. But in principle, after all, the process could be used to other ends. One could execute a living will directing that everything possible be done to keep oneself alive—even when one's "personal" capacities had been irretrievably lost. What then?

In a case somewhat like this, Helga Wanglie's caregivers answered that question by seeking a court order to stop the respirator and feeding tube that were sustaining her life. Mrs. Wanglie was an eighty-seven-year-old woman who, because of a respiratory attack, lost oxygen to her brain. She did not recover and remained in a persistent vegetative state. Although the costs of her care were covered by the family's insurance policy, the hospital still sought permission to remove life support. In some relatively minor ways, her case does not fit perfectly the hypothetical situation I considered above; for she had no living will. What she had, though, was a husband who was her guardian and who refused to consent to the withdrawal of treatment, believing she would not have wanted him to do so. And the medical caregivers went to court challenging her husband's suitability as guardian, rather than directly seeking court approval to terminate treatment.[20] But, as Alexander Morgan Capron notes, when they first announced their intention to go to court, the caregivers stated that "they did not 'want to give medical care they described as futile.' . . ."

Thus, in the Wanglie case, at least in the minds of the caregivers, personhood defined in terms of the right autonomously to determine one's future gave way to personhood defined in terms of the present possession of certain capacities.[21] For those who lack such rational capacities, further care is understood as futile—whatever they might previously have stipulated while competent. Similarly, when Schneiderman, Jecker, and Jonsen develop their "qualitative" understanding of futility, they make clear its impact on cases like this one. "The patient has no right to be sustained in a state in which he or she has no purpose other than mere vegetative survival; the physician has no obligation to offer this option or services to achieve it."[22] Ironies abound here. At the heart of the bioethics movement has been an assertion of personal autonomy for patients, which was, of course, ordinarily understood as ensuring their ability to be rid of unwanted treatment. But having built autonomy into the center of our understanding of personhood, having indeed (post *Roe v. Wade*) claimed that such autonomy flows from our right of privacy and may be asserted on our behalf even by others when we are unable to assert our wishes, having used patient autonomy as a hammer to bludgeon into submission paternalistic physicians, we suddenly rediscover the responsibility of physicians to consider what is really best for the patient, to make judgments about when care is futile. We suddenly do an about face. Against past autonomous patient choice for continued treatment even after "personhood" has been lost, we now assert medical responsibility not to provide present care that is "futile."

Helga Wanglie's caregivers, and those who would assert a "qualitative" notion of personhood are both right and wrong—though not in the ways they suppose. Right in that there is no reason to think that my physicians should forever be bound by what I stipulate (when I am forty-five and in good health) about my future care. (Right, that is, in thinking that autonomy alone is far too thin an account of the person—and that physicians must concern themselves with patients' best interests, not just their requests or directives.) Wrong in supposing that care for me becomes futile simply because I have irretrievably lost the higher human capacities for reasoning and self-awareness. But they are also confused; for the vision of the person guiding them where they are right is incompatible with the vision of the person at work where they are wrong. In supposing that care for me becomes futile when I have lost my powers of reason (even though I may not be terminally ill), they express a vision of the person that divorces personhood from organic bodily life. They decline to take into their understanding of the person defect, dependence, or disability. But in judging that caregivers need not be bound forever by directions I have stipulated in advance, when my condition was quite different than it has now become, they accept the need to live out the body's history, and they decline to give privileged status to the person's existence at one earlier moment in time.

If we could develop an increased sense of irony about the course the bioethics movement has taken, we might be well positioned to think about the important questions for everyday life with which it here deals. The ironies are a clue to our confusions. Is it not striking that—just at the moment when the idea of living wills seems to have triumphed, when federal law has required hospitals to make certain we know of our right to execute an advance directive—bioethicists should begin to wonder whether living wills are not themselves problematic? Having gotten what we thought we wanted—a law undergirded by a certain vision of the person—we begin to discern problems.

Thus, for example, John A. Robertson writes of "Second Thoughts on Living Wills."[23] There are, he notes, spheres of life in which we do not hold a person to an understanding he or she had previously stated. We do not, for example, hold surrogate mothers to contracts. Yet, we are reluctant to recognize that when Meilaender becomes incompetent—severely demented, let us say—his interests may well shift. We prefer to suppose that his person was complete and perfect at some earlier point in his development—when, say, at age forty-five he executed a living will. We hesitate to consider that what the forty-five-year-old Meilaender thought should be done to and for a demented Meilaender may not be in the latter's best interest. His life circumstances have changed drastically; he has become more simply and completely organism and less neocortex. If we would care for him, we must take that into account. And if we do not take it into our reckoning, if we blindly follow whatever directions the forty-five-year-old Meilaender gave, it is not clear that we can really claim to have the best interests of *this* patient—the Meilaender now before us—at the center of our concern.

In an argument similar in certain respects, Rebecca Dresser and Peter Whitehouse have suggested that our care for many demented patients should be grounded in an "objective approach" that "focuses on the incompetent patients' current condition (as opposed to prior preferences). . . ."[24] When we attempt to enter into the experience of demented patients, we will be moved neither always to treat nor always to withhold treatment. Thus, Dresser and Whitehouse seek to describe situations in which treatment itself (not just a condition of dementia) might constitute a significant burden for the patient and should not be imposed. But they argue just as strongly that for many demented patients treatment will not bring pain or distress, even though we may, of course, find their condition distressing. We must, therefore, be careful to distinguish our "personal discomfort about aging and mental decline from the patient's own subjective reality."[25]

These arguments make good sense. They essentially deny that we should think of the person as a perfect *eidos* captured at a moment in time, and, less directly, they invite us to think of the person as a someone who has a history, as animated earth. If, however, Dresser and Whitehouse aim to protect demented patients against hasty treatment withdrawals based on judgments about what they would, when competent, have desired, Robertson's aim is somewhat different. He sees that the living will has become essentially "a device that functions to avoid assessing incompetent patient interests." But his real aim is to encourage us to take up "the difficult task of determining which incompetent states of existence are worth protecting." This can only land him back in the muddle from which he is trying to escape. He is back to thinking of personhood as something added to existence—and well on his way, therefore, to the conception of personhood that gave rise to an emphasis on autonomy, which in turn suggested the living will as a useful way to exercise our autonomy, which—or so he thinks—is a path strewn with "conceptual frailties." He wants us not to live out the destiny of the body but to escape it.

IV

To have a life is to be *terra animata*, a living body whose natural history has a trajectory. It is to be a someone who has a history, not a someone with certain capacities or characteristics. In our history this understanding of the person was most fully developed when Christians had to make sense of the claim that in Jesus of Nazareth both divine and human natures were joined in one person.[26] Christians did not wish to say that there were really two persons (two sets of personal characteristics) in Christ; hence, they could not formulate his personal identity in terms of capacities or characteristics. They could speak of his person only as an individual with a history, a "someone who." The personal is not just an example of the universal form; rather, the general characteristics exist in and through the individual person. And we can come to know such persons only by entering into their history, by personal engagement and commitment to them, not by measuring them against an ideal of health or personhood.

Perhaps such an understanding of the person is also available to us through reflection upon our life as embodied beings. "Embodiment is a curse only for those who believe they deserve to be gods."[27] If Origen's account of the resurrected body seems to have lost much of what we mean by embodiment, he had at least this excuse: he genuinely believed that God intended to divinize humankind. That bioethics—and our culture more generally—is in danger of losing the body in search of the person is harder to understand, unless in our own way we believe that we deserve to be gods.

James Rachels, arguing that ethics can and must get along quite well without God, has recently distinguished between biological and biographical life, arguing that only the second of these is of any value to us.[28] The former means simply being organically alive; the latter means for him "having a life," having self-consciousness and self-control. Biological life has instrumental value, since apart from it there is no possibility of realizing biographical life, but biological life without the possibility of biography can be of no value to us. In such a state we no longer have any interest in living, and we cannot be harmed if our life is not preserved.[29]

Perhaps, though, such arguments do not take seriously enough the *terra* of which we are made. What Rachels never explains, for example, is why one's period of decline is not part of one's personal history, one's biography. As John Kleinig suggests, "Karen Ann Quinlan's biography did not end in 1975, when she became permanently comatose. It continued for another ten years. That was part of the tragedy of her life."[30] From zygote to irreversible coma, each life is a single personal history. We may, Kleinig notes, distinguish different points in this story—from potentiality to zenith to residuality. But the zenith is not the person. "Human beings are continuants, organisms with a history that extends beyond their immediate present, usually forward and backward. What has come to be seen as 'personhood,' a selected segment of that organismic trajectory, is connected to its earlier and later phases by a complex of factors—physical, social, psychological—that constitutes part of a single history."

Indeed, it is not at all strange to suggest that even the unaware living body has "interests." For the living body takes in nourishment and uses it, the living body struggles against infection and injury. And if we remember "the somatic dimensions of personality, as expressed for instance in face and hands," we may recognize in the living body the place—the only place—through which the person is present with us.[31] This does not mean that the person is "merely" body; indeed, in such contexts the word *merely* is always a dangerous word. As bodies we are located in time, space, and history; yet, we also transcend that location to some degree. Indeed, from the Christian perspective with which I began, it is right to say that, precisely because we are made for God, we indefinitely transcend our historical location. But it is as embodied creatures that we do so, and our person cannot be divorced from the body and its natural trajectory. This is not vitalism; it is "the wisdom of the body."[32] It is the wisdom to see that every human life is a story and has a narrative quality—a plot to be lived out. That story begins before we are conscious of it, and, for many of us, continues after we have lost consciousness of it. Yet, each narrative is the story of a "someone who"—someone who, as a living body, has a history.

Caught as we are within the midst of our own life stories, and unable as we are to grasp anyone else's story as a single whole, we have to admit that only God can see us as the persons we are—can catch the self and hold it still. What exactly we will be like when we are with God is, therefore, always beyond our capacity to say. But it will be the completion of the someone who we were and are—and we should not, therefore, settle for any more truncated vision of the person even here and now.

3. How Bioethics Lost the Body: Producing Children

When, quite near the birth of bioethics, the Hastings Center was founded in 1969 as the first bioethics "think tank," it planned research into four areas of concern: death and dying; behavior control; genetic screening, counseling, and engineering; and population policy and family planning. Although assisted reproduction is related to the third and fourth of these areas, it did not stand alone as a central concern. Nevertheless, it has received a great deal of bioethical attention in the quarter century since then.

One might argue that this attention is disproportionate to the number of people affected by advances in reproductive technology, especially when compared to the wide reach of some of the Center's other original areas of research. If, however, assisted reproduction quickly came to the fore, the reasons may be quite understandable. The *Roe v. Wade* decision in 1973 was wide ranging in its influence and in the kinds of questions it raised without answering. In addition, assisted reproduction—even if it touched relatively few people's lives directly—posed issues of great symbolic significance. It forced us to reflect upon the meaning of our common humanity. Such reflection has the capacity to reshape the way all of us live, even if we are not immediately affected by advances in reproductive technology. We sense quite rightly that the meaning of the human is implicated in such advance.

The publication in 1994, exactly a quarter century after the founding of the Hastings Center, of John Robertson's *Children of Choice* provides a suitable opportunity to reflect upon the fate of this topic in bioethical discourse.[1] Perhaps no author has written more widely than Robertson on this subject, the range of his influence is considerable, and this book might almost be described as a *Summa*. That he approaches these issues essentially as legal questions is itself an indication of a movement that has taken place in bioethics. In 1969 lawyers did not dominate bioethical discussions as clearly as they do today. And, in fact, although Robertson repeatedly suggests that he is discussing reproductive freedom as a moral and legal matter, I can find very little in his treatment that qualifies as moral argument. The wide-ranging humanistic concerns of Paul Ramsey's *Fabricated Man* (published in 1970) or, to take a very different normative position, Joseph Fletcher's *The Ethics of Genetic Control* (published in 1974) are largely absent from Robertson's treatment.[2] Whether this is progress we may doubt, but Robertson's argument deserves our attention precisely because it represents well a widely shared and influential viewpoint, the force of which is felt by many in our society. I will first explore the structure of his discussion and then reflect upon some of the deeper issues that he acknowledges but (with a rapidity that often astonishes) passes by.

I

Robertson takes up a number of different issues—abortion, forced contraception, in vitro fertilization and collaborative reproduction, the status of the human embryo, quality control of offspring, restrictions on pregnant women to prevent harm to children, and nonreproductive uses of our reproductive powers (e.g., producing embryos for research, conceiving a child to serve as an organ donor)—but throughout his discussion, the fundamental structure of argument remains essentially the same. He begins with the claim that (given the legal status of abortion in our society) there is a right *not* to reproduce. He then argues that it should follow that there is also a right to reproduce if one wishes. Although this would be, in the first instance, a right to reproduce coitally, Robertson argues that it must logically be extended to include noncoital reproduction as well as the power to control characteristics of the offspring one produces.

This right is basic but not absolute. Hence, it could be limited if the state had compelling interests that required such limitation. For Robertson, however, the only sort of interest that might warrant limiting the right of reproductive liberty would be clear evidence that harm would result from its exercise. Most of the objections raised against one or another manner of exercising the right seem to him not to point toward actual harms but to express "symbolic" concerns. Those concerns, he suggests, are matters about which citizens in a pluralistic society may reasonably differ, and, therefore, they offer no substantial ground for limiting the right of reproductive liberty. Even though relatively few limitations upon the exercise of this right are, in his view, justified, Robertson does describe it as a negative rather than a positive right. The state is not obligated to make possible our exercise of the right; it simply cannot stop us from exercising it if we wish. With that bare bones outline of the argument before us, we need to consider at least three of its important features: the meaning and scope of reproductive liberty, the importance of the liberty, and its character as a negative rather than a positive right.

THE MEANING AND SCOPE OF
REPRODUCTIVE LIBERTY

The goal of his book is, Robertson writes, "to show the importance of procreative liberty" (p. 4). Although this is the language Robertson often uses, I will generally refer to "reproductive liberty," since, as will become evident, I do not think he is discussing a phenomenon that can accurately be termed *procreation*. The first task, however, is to gain some clarity about the nature of the liberty.

Robertson's initial and simplest description of reproductive liberty is that it is "the freedom to decide whether or not to have offspring" (p. 4). Elsewhere he speaks of a "decision to have or not have children" (p. 5), and of "an individual or couple's choice to use technology to achieve reproductive goals" (p. 18). Slightly broader is a formulation that describes this liberty as "the freedom to have and rear offspring" (p. 119).

Robertson clearly describes reproductive liberty as "first and foremost an individual interest" (p. 22). That he often refers to the right of "couples" to exercise their reproductive liberty does not, I think, indicate anything other than a deferential nod in the direction of bourgeois morality. When in chapter 6 he takes up the question of collaborative reproduction (involving donors and surrogates), he raises at the very outset a hard question that has the capacity to undermine his individualistic description of reproductive liberty. Given that such collaborative undertakings involve "the decomposition of the usually unified aspects of reproduction into separate genetic, gestational, and social strands," we may ask: "Are couples [and even this is too narrow a term here, since the users of these techniques need not be a "couple" in any ordinary sense] who use these techniques 'procreating' in a significant way, even though one of them may lack a genetic or biological connection to offspring? Is a collaborator meaningfully procreating if he or she is merely providing gametes or gestation without any rearing role?" (p. 120). A reader will search the rest of this chapter in vain for any attempt to address that question. Instead, Robertson launches into a discussion of whether either the collaborators or the offspring are "harmed" by such reproduction. Since, apart from some merely "symbolic" concerns, such harm seems unlikely to him, he can find relatively little reason for limits on collaborative forms of reproduction. I suspect he thinks he has taken up the question he had raised: whether the parties to such an undertaking could meaningfully be said to be procreating. But, in fact, he has not come even within hailing distance of the question. To do so would require him to wrestle with some of those symbolic concerns he so regularly sets aside.

The exercise of reproductive liberty does not require any biological tie at all to the offspring produced. Recognizing the possibility of various collaborative arrangements in which people acquire a child to rear but have no biological connection to that child, Robertson suggests that, while this "is not reproduction in the strict sense," it still is part of reproductive freedom "because of the importance of parenting to persons who cannot reproduce themselves" (p. 143). And, to be even more precise, he should have written: "because of the importance of parenting to persons who cannot *or will not* reproduce themselves." Not only is noncoital, collaborative reproduction a part of reproductive liberty, so also are many measures one might take to control or shape the characteristics of the children one begets or intends to rear. Reproductive liberty is, Robertson asserts, of great personal significance, and "[i]f a person thought that she would realize those benefits only from a child with particular characteristics, then she should be free to select offspring to have those preferred traits" (pp. 152f.).

It turns out, in fact, that the connection of parent and child protected by the right of reproductive liberty is almost entirely a product of the will. Since collaborative reproduction is a part of this liberty, it cannot necessarily require an intention to rear the child who is the product of one's gametes. And, on the other hand, it cannot require any biological connection at all with the child whom one gestates and/or rears. Unpacking this liberty in detail leads Robertson to hypothesize that we will be led to reevaluate our understanding of the family (though this depends, of course, upon the unargued assumption that those who use such techniques are, in fact, doing the same thing as those who procreate in the traditional manner). "Such a reevaluation might show that preconception rearing intentions should count as much as or more than biological connection. If so, then arrangements in which several persons collaborate to produce a child for person(s) to rear who have no biological connection with the child should also be presumptively protected" (p. 143). This in turn suggests to him that a broadened understanding of the liberty involved may one day "lead to a widespread market in paid conception, pregnancies, and adoptions"—a possibility he views with equanimity (p. 143). What should be clear here is the triumph of will in Robertson's understanding of reproductive liberty. Whatever some people do from the current panoply of techniques to produce a child that one or more of them will rear seems presumptively protected by the right of reproductive liberty as Robertson understands it.

To be sure, there are some acts related to reproduction that are said not to fall within the scope of the liberty. Thus, for example, Robertson holds that creating embryos for research, while it ought to be permitted and protected for other reasons, cannot be defended by appeal to reproductive liberty (pp. 200f.). But having a child so that it can be a tissue donor *is* an exercise of this liberty, since it does, after all, involve the production of a child (p. 197). Only at "extreme measures such as cloning or nontherapeutic enhancement" does Robertson tend in the direction of prohibition (pp. 153f.). These measures, he says, "may violate widely shared notions of what makes procreation important" (p. 153). But a reader who has made it this far is likely to have difficulty explaining why. Robertson himself attempts to "posit a core view of the goals and values of reproduction" that will rule out those extreme measures (p. 167).

> On such a view, procreative liberty would protect only actions designed to enable a couple to have normal, healthy offspring whom they intend to rear. Actions that aim to produce offspring that are more than normal (enhancement), less than normal (Bladerunner), or replicas of other human genomes (cloning) would not fall within procreative liberty because they deviate too far from the experiences that make reproduction a valued experience. (P. 167)

I at least am unable to reconcile this statement with Robertson's more general views. Certainly on his view there is no reason to restrict the exercise of the liberty to a "couple." More generally, if people might be reluctant to exercise their reproductive liberty without the freedom to enhance the characteristics of their offspring, Robertson's standard mode of argument suggests that freedom to enhance would also become a part of the liberty. If a child who is molded to be in some ways less than normal would not, but for that act, have been born at all, Robertson's standard analysis again suggests that no harm has been done that child.[3] And if people wish for whatever reason to rear a child who is the replica of an already existing person, there is nothing in Robertson's depiction of reproductive liberty that should lead us to object. His appeal here to actions that "deviate too far from the experiences that make reproduction a valued experience" is a last-ditch attempt to find limits to a freedom that no longer presupposes any natural substratum and fails to pour meaning back into a concept that has become entirely the impoverished creature of human will.

THE IMPORTANCE OF
REPRODUCTIVE LIBERTY

Why should we think this liberty so important? When Robertson addresses that question, he tends to repeat a few formulations that serve to ease us past the deeper humanistic issues that lie buried in such a question. Most generally, he suggests that certain "reproductive experiences . . . are central to personal conceptions of meaning and identity." They provide "a crucial self-defining experience" (p. 4). Or, again, the achievement of reproductive goals is "a central aspect of people's freedom to define themselves through reproduction" (p. 18).

Interestingly, however, not all reproductive experiences that might be termed self-defining are viewed by Robertson as part of reproductive liberty itself. Thus, for example, "whether the father may be present during childbirth, . . . or whether childbirth may occur at home rather than in a hospital may be important for the parties involved, but they do not implicate the freedom to reproduce (unless one could show that the place or mode of birth would determine whether birth occurs at all)" (p. 23). There is, of course, something quite sensible about such a claim, but its good sense stands in tension with the expansiveness and narcissism of reproductive experience as Robertson describes it. What is peripheral to one person's self-defining experience may be quite central to another's, and I cannot find a thread in Robertson's argument strong enough to bear the weight of these distinctions. Once we begin to appeal to the importance of certain private experiences for personal conceptions of meaning and dignity, it will not be easy to find our way back into worlds of shared meaning. Robertson hopes, however, that his second chapter will show that "procreative liberty deserves presumptive respect because of its central importance to individual meaning, dignity, and identity" (p. 16).

When we turn to that chapter, we are told that "transmission of one's genes through reproduction is an animal or species urge closely linked to the sex drive" (p. 24). Moreover, connection to future generations gives us solace in the face of death, may be the "expression of a couple's love or unity," and may have "religious significance" (p. 24). This is, I think, as close as Robertson comes to helping us understand the human importance of reproductive experience. Consider the possibility of a married couple seeking donor insemination because of the husband's infertility. The resulting child cannot be said without considerable argument to express that couple's unity, nor, of course, does the birth of a child involve transmission of the husband's genes. To suppose that the sperm donor is himself fulfilling that fundamental urge to transmit one's genes must entail that we think of the donor as personally present in the child—in which case donor anonymity becomes morally suspect. Thus, many of the considerations Robertson mentions here as reasons for the importance of reproductive experience are not involved in a common form of assisted reproduction. We are left with the solace that a possibly very attenuated link to future generations may offer us in the face of death, though we are not told exactly why giving rise to those who will replace us—and who must therefore remind us of our mortality—will provide such solace.

Moreover, even these considerations are sometimes interpreted by Robertson in ways that drain them of much of their ordinary significance. Thus, for example, if one has already given rise to offspring, the "marginal value" of additional offspring may be diminished (p. 31). Or, again, in claiming that people unfit to be parents need not be thought of as losing their right to reproductive liberty, Robertson notes that they might reproduce without rearing. "Offspring could be protected by having others rear them without interfering with parental reproduction" (p. 31). True as this is—and necessary as it unfortunately is in many cases—it suggests that we have not gotten too far in plumbing the human meaning of procreation.

For the most part, then, we are left with generalities about "the centrality of reproduction to personal identity, meaning, and dignity" (p. 30). We are to understand as centrally involved in the dignity of persons their "wish to replicate themselves" (p. 32). Robertson never asks in this connection whether human dignity might best be displayed in the way we deal with what is unwanted and unexpected in life. He never takes up rigorously the question whether those who wish to experience only biological parenthood, only rearing, or only gestating, are *doing* the same thing as those who hold these aspects of parenthood together—and whether, therefore, the importance we ascribe to the experience of parenthood is rightly conferred upon those who deliberately separate its constituent parts and seek to experience only some of them. He never asks whether bearing and rearing children is better thought of as a task or as a return we make for the gift of life than as an experience sought for purposes of self-definition—or whether we would ascribe to it the importance we do if we thought of it chiefly as an exercise in self-definition. In short, the reader is repeatedly assured that reproductive experience is of immense human importance, but the argument itself threatens to drain from the experience most of what has made it seem important.

A NEGATIVE RIGHT

Not only is reproductive liberty first and foremost an *individual* interest, it is also a negative rather than a positive right (p. 23). That is, although others may not interfere with one's exercise of the right, they are under no obligation to provide the services or resources that would make such exercise possible. Nevertheless, although the state is not obligated to make exercise of the right financially possible, it turns out that, on Robertson's analysis, it must do a good bit to foster such exercise.

Thus, for example, the right must include permission to exercise "quality control" over one's offspring; for apart from "some guarantee or protection against the risk of handicapped children," people might not choose to reproduce (p. 33). Or again: "Selection decisions are essential to procreative liberty because of the importance of expected outcome to whether a couple will start or continue a pregnancy" (p. 152). Evidently, then, the state could not seek to discourage such quality control without infringing upon the right to reproductive liberty. More puzzling still is a statement Robertson makes in his discussion of the contraceptive Norplant. "If women are to be guaranteed control over their fertility through contraception, long-acting contraceptives such as Norplant should be made available to all women who desire it" (pp. 70f.). Perhaps Robertson intends here only to make a policy recommendation and not to claim that such access to Norplant must be construed as part of the right of reproductive liberty; however, the language he uses suggests that he is discussing something that is part of the decision whether or not to have offspring—and, hence, part of the protected liberty. And this, in turn, sounds far more like a positive than a negative right.

Similarly, preconception agreements that are a part of some collaborative reproductions should, he argues, be regarded as binding, since people who cannot rely on such agreements will lack "the assurance they need to go forward with the collaborative enterprise" (p. 126). We are told that agreements to pay surrogates are "probably necessary if infertile couples are to obtain surrogacy services" (p. 140). Once again it seems that any restrictions designed to discourage the reduction of procreation to contractual terms are interpreted as infringements of the right. Despite this, Robertson writes that the state is not required to "subsidize or otherwise encourage the use of all reproductive techniques" and that states may even "refuse to enact laws that facilitate collaborative reproduction" (p. 234).

Considerably more clarity is needed about what it means to characterize reproductive liberty as a negative rather than a positive right. I suspect that the source of much of the unclarity is a general blurring of the moral and the legal throughout the argument. Perhaps Robertson, while recognizing that states may not presently be required under law to encourage assisted reproduction and that, indeed, they may even discourage it, nevertheless himself wishes to propose that such encouragement be brought within the scope of reproductive liberty. That, however, might involve the transformation of a negative right into a positive right.

II

Thus far I have focused chiefly on Robertson's discussion of the desire to reproduce. In fact, however, it is the desire *not* to reproduce that is more foundational in his argument. This is so for the quite simple reason that his argument is almost exclusively a legal (rather than moral) one, and it is the wish not to reproduce that is more firmly grounded in our present understanding of constitutional law. It is, therefore, worth paying some heed to Robertson's discussion (in chapter 3 of his book) of abortion.

The chapter is devoted largely to a discussion of what Robertson terms "a modified pro-choice position that is likely to dominate ethical, legal, and popular thinking about abortion for the foreseeable future" (p. 45). Here as elsewhere in the book a reader may not always be sure whether Robertson is describing or affirming, but, in general, he seems to look favorably upon this modified pro-choice position. From that perspective "abortion at early stages of pregnancy is generally viewed in most circumstances to be an ethically and legally acceptable act, but an act that should be discouraged or avoided whenever possible" (p. 46). If, however, Robertson does in fact support this position, he has a peculiar understanding of what it would mean to discourage an act. He recommends, for example, that in order to overcome distributive inequities in access to abortion, our public policy should "fund or provide contraceptive and abortion services" (p. 48). And, he writes, the acceptability of waiting periods longer than twenty-four hours for an abortion would depend on whether "such a period of reflection actually aids an informed decision or is merely obstructionist" (p. 62)—a peculiarly negative way of describing a policy that might reasonably be designed to discourage abortion.

Arguments in defense of abortion generally take one of two forms.[4] "Personhood" arguments justify abortion up to that point (if any) in pregnancy at which the fetus is thought to have become a person with rights. "Bodily support" arguments rely on the claim that, even if the fetus has rights, a pregnant woman cannot be obligated to provide it with the support of her body. The second of these arguments is principally a right not to have to carry a fetus—which is not the same as the right to a dead fetus. Only by the accidents of medical technology will its exercise in the early stages of pregnancy necessarily result in a dead fetus. That second argument, therefore, is more difficult to relate to Robertson's right of reproductive liberty, which is described as involving a decision to have or not have children. Obviously, even if one is not compelled to carry a fetus to term or rear the child who is born, one may still have a child if abortion is not understood necessarily to result in a dead fetus. (And, as Robertson notes, "[e]ven if the child is relinquished for adoption, there will be powerful feelings of attachment, responsibility, and guilt that will, in many cases, last a lifetime" [p. 49]). Hence, one might expect that the personhood argument would carry more weight for Robertson. And he says, in fact, that abortion is about "escaping those burdens" not just of carrying or caring for a child but of having one's offspring alive against one's will (p. 49).

Robertson faces the difficulty, however, as he clearly recognizes, that the Supreme Court's reasoning has depended less on the personhood than on the bodily support argument. By making viability a crucial, if not fully determinative, line, the *Roe v. Wade* decision ascribed legal (and perhaps moral) significance to the point at which the fetus can survive outside the womb without the mother's bodily support (p. 54). And, as Robertson also notes, if the fetus is considered to have personal status, the woman's "morally protectable interest . . . [would consist] in becoming free of bodily burdens and not in avoiding reproduction altogether" (p. 51). Given his understanding of reproductive liberty, it is not surprising that Robertson should seek to move the argument in a different direction.

The importance of viability, he suggests, is not that it marks the point at which the fetus can live without the mother's bodily support. Rather, it is around the time of viability that the fetus becomes sentient "and thus has interests in its own right" (p. 53). This does not mean that the fetus is yet a person with rights, since personhood requires "the ability to reason or make choices" (p. 51). But once the fetus is sentient, we may have some moral duties toward it as we do toward animals (p. 53). For the moment, then, as long as viability roughly coincides with the appearance of sentience, it will be a morally significant line (as it is in *Roe v. Wade*). But the crucial point Robertson sees and states clearly: "If technology pushes viability back to earlier presentient stages, it will cease to have this moral significance, because survivability will no longer correlate with sentience" (p. 53).

To rest too much weight on the bodily support argument would endanger the entire structure of Robertson's case for reproductive liberty. For if the day comes, as it well may, when fetuses can be kept alive outside the womb even before they are sentient, the reasoning of *Roe v. Wade* would no longer under-gird a right to abortion. Thus, Robertson needs the broader right not to have children, the right to a dead fetus, and that in turn will require some version of a personhood argument to suggest that we do no injustice when we abort the fetus. Here again, of course, one is never certain whether Robertson is pressing a moral argument or predicting (or attempting to shape) the future course of constitutional interpretation. While he seems generally to suggest that he is doing the latter, I think it is the former—a largely undefended argument about the meaning and moral importance of personhood—that is driving the argument.[5]

He claims, for instance, that a blanket condemnation of all abortion after conception "overlooks the very different bio-logic stages of embryonic and fetal development" (p. 48). But his claim is underdeveloped in several ways. Once we have made the personhood—and not the bodily support—argument central, it will readily occur to us that an individual human life goes through a variety of developmental stages both before and after birth (a moment that has moral significance accord-ing to the bodily support, but not the personhood, argument). We may therefore be uncertain why Robertson says that "a person's keen interest in avoiding the social burdens of re-production does not justify infanticide" (p. 50). Indeed, he has given us no argument that supports that claim. We may further note that the process of development within a human life generally follows a trajectory that includes, at the end, de-cline—sometimes decline into a condition in which one lacks the ability to reason or make choices, sometimes into non-sentience. We will have to consider what the full implications of Robertson's understanding of personhood are before we can decide whether it represents the way of wisdom.

In short, although Robertson writes of "the extremes and inconsistencies of the anti-choice program," (p. 66), his own proabortion stance is extreme (in its reduction of human per-

sonal dignity to cognitive and volitional capacities) and in-
consistent (in its failure to follow through on his preference for
the personhood argument). He detects in pro-lifers a "latent
agenda" that sees in abortion a denigration of the importance
of sex and marriage and "an attack on or devaluation of their
life-style" (p. 67). One would be more impressed with his capacity to discern such agendas, though, were he to note that the
interest of many men in their sexual freedom leads them to
espouse pro-choice views, or were he to consider whether
some pro-choicers might themselves feel threatened by a moral
ideal that gives compelling testimony to human interdepen-
dence and the strength of character required to deal with the
unexpected and unwanted in life. To explore such questions,
not just to detect latent agendas, is the true task of moral reasoning.

III

At the very outset of his discussion, on the first page of the
introductory chapter, Robertson declares that there is "something profoundly frightening" about the forms of technological advance and kinds of choices that he will be discussing
(p. 3). And at the end of the book he writes of our "ambivalence"—both individual and societal—toward these techniques. But I suspect that a reader of the intervening pages
would probably not discern much ambivalence or worry in
Robertson's voice. It is, therefore, a little difficult to take to
heart professions of ambivalence unless we bring other voices
into play. That will be my aim here. What Robertson offers us is
a relatively straightforward but very thin understanding of
human life. Individuals are largely isolated wills, brought together in association when they choose to cooperate in pursuit
of their interests. Such an intense but narrow focus might

remind us of C. S. Lewis's contrast between the depiction of Adam and of Satan in Milton's *Paradise Lost:*

> Adam, though locally confined to a small park on a small planet, has interests that embrace 'all the choir of heaven and all the furniture of earth'. Satan has been in the Heaven of Heavens and in the abyss of Hell, and surveyed all that lies between them, and in that whole immensity has found only one thing that interests Satan. . . . [His] monomaniac concern with himself and his supposed rights and wrongs is a necessity of the Satanic predicament. . . . He has wished to 'be himself', and to be in himself and for himself, and his wish has been granted.[6]

A single-minded focus on the self's willful pursuit of its projects in the world has the effect, it turns out, of obscuring our vision of other important realities. To attend to these other considerations we turn now to develop a vision of the individual as more situated and embedded—and to consider whether we may not thereby see more deeply into the meaning of human procreation.

MARRIAGE AS A BASIC FORM OF HUMAN LIFE

Robertson recognizes, as I noted earlier, that those who decompose marriage into its constituent parts, who separate its relational and procreative dimensions, may not be "doing" the same thing as those who beget and rear children within the bond of marriage. Despite recognizing the legitimacy of such a question, however, he never fully addresses it. It is therefore worth our seeking to understand why one might believe that the union of biological, gestational, and rearing parenthood within a marriage should be important for human life.

Certainly we are able to separate the personal and the biological, the relational and the procreative, dimensions of marriage and then recombine them in a variety of ways. That we can do so testifies to the marvelous range of human freedom. If such freedom is the sole truth about human nature, if we are simply beings who freely create ourselves, there will be no limit to such self-definition and self-creation other than the limit of harm to others, which Robertson regularly invokes. Human dominion over the natural world has in recent years often been seen as problematic. But in our schizophrenic culture we are able to deplore environmental abuse as unwarranted exercise of human freedom while forging ahead in turning procreation into reproduction. Rather than seeing the person as present in the body, and the person's freedom as necessarily dependent upon respect for the natural world, we suppose that the personal and the biological are entirely separate realms and that we need not honor the body as the locus of personal presence.

Clearly, there is given in human nature a connection between procreation and the differentiation of the sexes. "To this given connection in our nature . . . it is possible to respond in only two ways. We may welcome it, or we may resent it."[7] We may resent it and use our freedom to remake it in the countless different ways that Robertson discusses—and in more ways still to be imagined. We may, that is, find human meaning in the body's procreative powers only when we choose that it should be there and only in the manner that we choose. But we may also honor the body as the presence of the person and ask ourselves what the good might be of holding together procreation and the bond of personal love between a man and woman.

Understanding procreation as appropriate only within the bond of mutual love of husband and wife will, first, be good for the loving relationship itself. No doubt the motives of those who beget children coitally within marriage are often mixed, and they may have much to learn about the meaning of their action. But if they are willing to shape their intentions in

accord with the nature of the act itself, in learning what procreation means they may be freed from self-absorption. Rather
than being an exercise in self-definition or self-replication, procreation, as the fruition of coitus, should teach us that the act
of love is not simply a personal project undertaken for our own
fulfillment. That the embrace of husband and wife may prove
fruitful and may sustain human life can give to their love a
spaciousness it needs. Even when the relation of a man and
woman does not or cannot give rise to offspring, they can
understand their embrace as more than their personal project
in the world, as their participation in a form of life that carries
its own inner meaning and has its *telos* established in nature.
Some understanding like this is needed if the sexual relation of
man and woman is to be more than "simply a profound form
of play. . . ."[8] That our culture desperately needs to reclaim
such a vision before the relation of men and women is debased
still further seems, to me at least, evident.

Understanding procreation as appropriate only within the
bond of mutual love of male and female will, second, be good
for procreation itself. Even when a man and woman deeply
desire a child, even when their act of love is moved by that
desire and hope, the child can never be "the primary object of
attention in that embrace."[9] Indeed, unless one's beloved is the
object of one's attention and desire, the man, at least, may even
be unable to fulfill his designated role. In the act of love the
partners must set aside their projects in order to give themselves to each other, and the (perhaps hoped for) child becomes
the natural fruition of their shared love, not a chosen project.
The child, therefore, is always a gift and, even, a mystery—one
like them who springs from their embrace, not an inferior
being whom they have made and whose destiny they should
now determine. That our culture desperately needs to reclaim
such a vision before our care for children is debased still further
seems, to me at least, evident.

It is clear, of course, that we are able to sever the procreative and relational dimensions of marriage at will, to find no personal signficance in their natural unity unless we choose that it should be there. In our freedom we can thus soar far above our finite condition, forgetting that there may be more ways to violate our humanity than by limiting that freedom. Certainly, we ought not suppose that an affirmation of reproductive liberty comes unencumbered with any metaphysical baggage. For, as Paul Ramsey once noted, when we dismember procreation into its several parts and combine them in new and different ways, we simply enact a new myth of creation in which human beings are created with two separate faculties—one manifesting the deepening of the unity of the partners through sexual relations, the other giving rise to children through "a cool, deliberate act of man's rational will."[10] Thus, a symbolic vision of human nature is present both in the position I have outlined above *and* in Robertson's defense of reproductive liberty. Neither can be offered free of metaphysical implications. We should not therefore assume that those who "procreate" and those who—having severed procreation into its parts—"reproduce" are *doing* the same thing. But if we can learn again to think of marriage as a basic form of life within which procreation ought to take place, we will of necessity turn away from the direction of Robertson's argument at any number of important points. We will characterize procreation not simply as a right, but as the internal fruition of the act of love—a task to be taken up for the sustaining of human life. We will see all forms of collaborative reproduction as dehumanizing, as a violation of a basic form of human life. We will doubt whether "quality control" of our offspring can express a commitment to human equality that envisions the child not as a product we have made but as one like us in dignity.

MAKING AND DOING

This last point—the child as one who is begotten, not made—deserves more extended development. The contrast between doing and making is at least as old as Aristotle's characterization of *praxis* and *poesis*.[11] This distinction has had no impact on Robertson's discussion, however. When, for example, he discusses the permissibility of noncoital techniques for reproduction, he notes that some people have moral objections. "However, often the harms feared are deontological in character. In some cases they stem from a religious or moral conception of the unity of sex and reproduction or the definition of the family" (p. 34). One is tempted to respond: "Well, yes. What sort of objections do we think we might encounter when contemplating the possibility of moral concerns?" But the point is that, for Robertson, making rather than doing, what we accomplish rather than what we do, always wins the argument. "[W]ithout a clear showing of substantial harm to the tangible interests of others, speculation or mere moral objections alone should not override the moral right of infertile couples to use those techniques to form families" (p. 35). Why then, one might ask, did he bother even to mention moral objections? "Given the primacy of procreative liberty," he writes, "the use of these techniques should be judged by the same restrictive standard that would be applied to coital reproduction" (p. 35). The point, however, is that—at least if Robertson takes himself to be engaging in moral argument—this is not "given." He assumes and asserts it, but largely without moral argument. And in so doing he narrows tremendously the range of considerations that may shape our thinking about the meaning of the presence of children in human life. We need to open our minds to other features of human action in order to see how we might come to think of a child as our equal—and not simply as our product or project.

In passing quickly by any objections thought to be "deon-tological" in character, Robertson is declining to think about what it means to "do" something, refusing to contemplate the difference between what we do and what we accomplish in our doing. A richer understanding of the ways of evaluating human behavior will make clear just how much he omits. In *The Responsible Self* H.Richard Niebuhr delineated three dif-ferent modes of moral reasoning and evaluation.[12] I will draw on his categories here to suggest how much is missing in Robertson's treatment.

We may describe human beings, first, as makers or fash-ioners. We act because we have goals and want to realize goods that we value. We are in the world as people who have projects, which we seek to realize through action. Like artisans or crafts-men, we are at work on a product, attempting to fashion it into the desired shape. From this perspective we necessarily pay at-tention chiefly to what we accomplish, to the results of our action. Seeking to enhance human life as much as we are able, we cannot help but count the cost—weigh costs and benefits—when pursuing our projects and shaping our products. When we think in this way, our attention focused on the goal we seek to realize, we quite naturally suppose that an "ought to do" follows from an "ought to be." If it ought to be the case that people experience the sense of worth and dignity that comes from producing children, then they and we ought to do what-ever makes possible such significant experience for them—unless, of course, the harm to others that comes of this is too great. That this first method of moral evaluation is important we cannot doubt, since we are indeed present in the world as goal-oriented, productive beings. True as this is, however, there is more to be said about human action.

Second in Niebuhr's typology is the fact that we come to know ourselves as human agents only within a community of others in whom we recognize a being and dignity like our own.

> The sociality of the self refers to the fact that our very possibility
> of being human subjects comes to us through our relationships
> with significant others. I need the concern and positive regard
> of others in order to actualize my own personhood. This need
> occasions the disclosure of the value others hold for me. Yet I
> discover a new fact in these need-fulfilling relationships: these
> "others" are in a position to lay claims upon me.[13]

Put less abstractly, in recognizing my own dignity I recognize
the dignity of others like me. In recognizing that I can sometimes be wronged by others even without being harmed by
them, and that they likewise may be wronged by me even without being harmed, I come to see that our humanity is
expressed in what we do, not just in what we make or accomplish. We may have obligations to do or refrain from doing that
are not grounded in harms to others. Hence, on this model
Niebuhr pictures human beings not as makers or fashioners but
as fellow citizens—sharing a common life governed by rules
that shape action in ways appropriate to their equal dignity.
Recognizing therefore the complexity of human action, that
there is more to be said about it than just that it is goal oriented, we may also learn to contemplate the possibility that
some things should not be done even if they would achieve results that might be, on the whole, desirable. This is not simply
a "deontological" concern; it is integral to any serious consideration of human action.

If human action is goal oriented and intersubjective, it is
also—and third in Niebuhr's typology—responsive to all that
presses upon us and claims our attention. In Niebuhr's terms,
we seek therefore to act not only in ways that realize what is
good or do what is right, but also in ways that are "fitting"—
that fittingly respond to all that is acting upon us and shaping
us. Just as being a good driver involves not only knowing
where we are going and being familiar with the rules of the

road but also responding in countless ways to what is going on around us, so also the discerning moral agent must understand his action as limited and responsive.[14] If, with Robertson, we restrict moral considerations to goals and values, we see human beings only as makers, engineers. We lose then an understanding of community in which our action is limited by the equal dignity of our fellow citizens. And we lose the sense that we are not, finally, beings characterized only by self-modifying freedom. We are also responders: to nature, to others like ourselves, and to what transcends both nature and humanity.

However important "making" may be, therefore, it does not exhaust the categories by which we should think about and evaluate human action. Moreover, there may be occasions when it is an inappropriate category, when it cannot capture the human signficance of what we do. Procreation is such an occasion, for only the child who is "begotten, not made" can be one equal to us in dignity, one who is not finally a product at our disposal.[15] We must think of the body as the locus of personal presence in order to discern the equal worth of the child who springs from the embrace of our bodies. There are countless ways to "have" a child. Not all of them amount to doing the same thing. Not all of them will teach us to discern the equal humanity of the child as one who is *not* our product but, rather, the natural development of shared love, like to us in dignity.

The formation of a family is most truly human when it springs from what Gabriel Marcel called "an experience of plenitude."[16] To conceive, bear, give birth to, and rear a child ought to be an affirmation and a recognition: affirmation of the good of life that we ourselves were given; recognition that this life bears its own creative power to which we should be faithful. In this sense Marcel could claim that "the truest fidelity is creative."[17] That something rather than nothing exists is a mystery lying buried in the heart of God, whose creative power and plenitude of being are the ground of our life. That a new human being should come into existence is not ultimately our doing. Within this life we can exercise a modest degree of control, but if we seek to do more than that we have fundamentally altered the nature of what we are doing—and of the beings to whom we give rise. Therefore, to form a family ought not be an act of planning and control by which we replicate ourselves or gain access to a pleasurable experience of our own worth. It ought to be an act of faith and hope, what Marcel termed "the exercise of a fundamental generosity."[18]

SYMBOLS

Summing up his argument, Robertson at one point writes: "The invocation of procreative liberty as a dominant value is not intended to demolish opposition or end discussion. . . . Procreative choices that clearly harm the tangible interests of others are subject to regulation or even prohibition" (p. 221). Yet, as we have seen, Robertson's discussion is in fact deaf to the wide range of considerations included in moral reflection. All these he tends to set aside as merely "deontological" concerns, setting aside thereby much that is central to our humanity. But the really "magical" word in Robertson's book, the word that truly does end discussion, is *symbolic*. Summing up again near the end of the book he writes: "As illustrated repeatedly throughout this book, many of the concerns and fears will, upon closer analysis, turn out to be speculative fears or symbolic perceptions that do not justify infringing core procreative interests" (p. 222). It is necessary but insufficient to point out that this has not been "illustrated," much less demonstrated; it has only been asserted.

But it has been asserted repeatedly. In outlining the course his argument would take, Robertson notes at the outset that the interest in reproductive liberty, though not absolute, will often trump "competing concerns that are too speculative or symbolic to justify intrusion on procreative choice" (p. 17). The view that it is wrong to separate reproduction from the marital bond he describes as "symbolic," a matter over which reasonable people may differ (p. 41). Indeed, "concerns about the decomposition of parenthood through the use of donors and surrogates, about the temporal alteration of conception, gestation and birth, about the alienation or commercialization of gestational capacity, and about selection and control of offspring characteristics" do not involve "substantial harm to tangible interests of others" but affect only our "notions of right behavior" (p. 41). Discussing abortion, he asserts that "moral objections or symbolic commitments alone, over which individuals in a pluralistic society usually make their own choice" cannot override the right not to bear offspring (p. 50). The unwillingness of some people to view as binding a pre-conception agreement between collaborators in reproduction is, Robertson thinks, "based on paternalistic attitudes toward women or on a symbolic view of maternal gestation" (p. 132). Objections to a market for the sale of gestational services express a "symbolic concern" about which "reasonable people have differing moral perceptions" (p. 141). Many objections to nonreproductive uses of our reproductive capacity involve "largely symbolic moral claims on behalf of embryos, fetuses, offspring, and women" (p. 198). Abortion of an existing pregnancy for transplant purposes might be reasonable, since one might well think that "the additional symbolic devaluation of human life through deliberate creation and destruction of pre-natal life is negligible" (p. 214). Clearly, the word *symbolic* functions as a mantra throughout the discussion, its very invocation seeming to settle disputes.

But, as Paul Tillich might have put it, we should never say "only a symbol." One of the important characteristics of a symbol is that "it opens up levels of reality which otherwise are closed for us."[19] Still more, symbols do not express thoughts that we (privately) have; they give rise to thought.[20] We cannot think in nonsymbolic ways. Indeed, even the most simple and least mythic human attempts at speech will be freighted with symbolic expression.[21] We may make our language less interesting; we cannot make it nonsymbolic. We can only be tone deaf to the symbols, as Robertson is, for example, when he reads Judith Jarvis Thomson's defense of abortion. In her hypothetical cases the fetus is symbolically construed as a parasite (in the unconscious violinist analogy) and as a mushroom (in the people-seed analogy), but all this seems quite straightforward to Robertson.[22] According to Robertson, Thomson "shows that in many cases, such as rape and sexual intercourse with contraception, a persuasive moral claim that the fetus has the right to use the body of another cannot be made" (p. 51). But Thomson's is not straightforward language about rights and interests. Embedded in her language are symbols of the human. To miss them is to misunderstand the argument.

Because symbols give rise to thought and expression, we overlook them or dismiss them at the peril of our humanity. If we think of them as "mere" symbols, we cut ourselves off from much that is most important in the life of human beings who are, after all, the symbol-making animals. Worrying precisely about such possibilities, C. S. Lewis once described the person who attempts to see through the symbolic nature of our language rather than seeing with it.

> Quite truly, therefore, he claims to have seen all the facts. There *is* nothing else there; except the meaning. He is therefore, as regards the matter in hand, in the position of an animal. You will have noticed that most dogs cannot understand *pointing*. You point to a bit of food on the floor: the dog, instead of looking at the floor, sniffs at your finger. A finger is a finger to him, and that is all. His world is all fact and no meaning.[23]

I have suggested above that the transformation of procreation into reproduction involves us in new ways of thinking about human life—which, of course, is not surprising, since symbols give rise to thought. Perhaps most dangerous is the possibility that we will find it more difficult to think of the child as one who is equal in dignity to those who make it. It is true, of course, that for some time to come our inherited ways of thinking may encourage us to think of children as equal to those who produce them. "But if we do not live and act in accordance with such conceptions, and if society welcomes more and more institutions and practices which implicitly deny them, then they will soon appear to be merely sentimental [that is, "merely symbolic"], the tatters and shreds which remind us of how we used once to clothe the world with intelligibility."[24]

When we think of human beings chiefly as "will," as beings characterized by their interests, we see something true, but we miss much else. We miss ways in which, subjecting the body to their will, they may endanger their humanity, threaten their equal dignity, and degrade their status as the symbol-making animal. Learning to think of human beings as will and freedom alone has been the long and steady project of modernity. At least since Kant, ethics has often turned to the human will as the sole source of value. The understanding of reproduction that Robertson depicts and defends is, from this perspective, not at all surprising; for it is faithful to that very narrow understanding of the human. If we are not surprised by it, however, that itself may be cause for worry. We need to think about more than arguments in law when we read Robertson's *Summa*. We need also to contemplate the image of our humanity symbolically portrayed in his pages.

4. Bioethics as Public Policy: A Case Study

In the first chapter I suggested that a useful bioethics would be one that no longer understood its purpose to be achieving social consensus. Surveying the manner in which both "principlism" and casuistry had become methods aimed at the development of public policy, I noted that in so doing bioethics had lost the "soul"—lost concern for what Leon Kass called "the deepest matters of our humanity" that had originally "animated the enterprise." In chapters 2 and 3 I have taken up some questions that arise at the end and at the beginning of life in order to illustrate how this loss of metaphysical and religious substance has meant a loss of the meaning of embodied personhood. Now I turn to one recent example of the poverty of bioethics: the Report of the Human Embryo Research Panel, established by the Advisory Committee to the Director of the National Institutes of Health.[1]

I

This report does not mark the first time such an issue has been a matter of public debate, and it is worth recalling earlier debates that set the context for this one. Prior to the establishment of the Human Embryo Research Panel, there had been two occasions in the last several decades in which publicly appointed committees had studied similar questions. After the *Roe v. Wade* decision in 1973, with the potential availability of many more (aborted) fetuses as research subjects, fetal research became a hotly debated issue. Congress established the National Commission for the Protection of Human Subjects of Biomedical and Behavioral Research and charged it to give attention first to the ethics of fetal research. The Commission published the results of its deliberations in 1975.[2] With certain safeguards it permitted research on the fetus *in utero* and upon the possibly viable infant outside the uterus.[3] Most important of the safeguards is a stipulation that any research not aimed at benefiting the fetus who is the research subject must impose "minimal or no risk" or "no additional risk" to the well-being of the research subject. But a harder question concerned research on the fetus still in the womb but intended for abortion, or research on the still living but *non*viable fetus outside the womb after abortion. At first sight the Commission applied the same standard to this research. It too should pose no more than minimal risk to the research subject.[4]

The difficulty of separating the issue of fetal research from that of abortion becomes apparent, however, when we consider the intricacies of the Commission's position at this point. Noting that equal respect need not always require identical treatment, the Commission contemplated the difference abortion makes. It is worth quoting several paragraphs from the "Deliberations and Conclusions" of its Report.

> The Commission affirms that the woman's decision for abortion does not, in itself, change the status of the fetus for purposes of protection. Thus, the same principles apply whether or not abortion is contemplated; in both cases, only minimal risk is acceptable.

Differences of opinion have arisen in the Commission, however, regarding the interpretation of risk to the fetus-to-be-aborted and thus whether some experiments that would not be permissible on a fetus-going-to-term might be permissible on a fetus-to-be-aborted. Some members hold that no procedures should be applied to a fetus-to-be-aborted that would not be applied to a fetus-going-to-term. . . . Others argue that, while a woman's decision for abortion does not change the status of the fetus per se, it does make a significant difference in one respect—namely in the risk of harm to the fetus. For example, the injection of a drug which crosses the placenta may not injure the fetus which is aborted within two weeks of injection, where it might injure the fetus two months after injection. . . .

There is basic agreement among Commission members as to the validity of the equality principle. There is disagreement as to its application to individual fetuses and classes of fetuses. Anticipating that differences of interpretation will arise over the application of the basic principles of equality and the determination of "minimal risk," the Commission recommends review at the national level.[5]

This issue was further clarified in an additional Statement by one of the Commissioners, Karen Lebacqz, with which another Commissioner, Albert Jonsen, concurred. She noted that to speak of "minimal risk" of harm required that some notion of "harm" be at work. This must, she thought, involve either (1) injury or diminished faculty, or (2) pain. Accepting the view that seemed to have been adopted by the Commission that the fetus cannot feel pain, and considering that a fetus destined for abortion or already aborted had no future life expectations to be further diminished, she wrote: "[I]n a dying subject prior to viability, 'diminution of faculties' does not appear to be a meaningful index of harm since this index refers largely to future life expectations." In short, if the fetus could not feel pain, and if its future life expectations could no longer be harmed, then "minimal risk of harm" might mean something quite different for it than for a fetus-going-to-term. The choice of abortion could open up possibilities for research that would never be permitted on a fetus not destined for abortion. Precisely such a possibility led dissenting Commissioner David Louisell to write: "Although the Commission uses adroit language to minimize the appearance of violating standard norms, no facile verbal formula can avoid the reality that under these Recommendations the fetus and nonviable infant will be subjected to nontherapeutic research from which other humans are protected."[6] Louisell was right to worry about the claim that equal respect could so easily be reconciled with unequal treatment; for here the difference in life prospects of potential research subjects is entirely a product of our own will and choice. This makes it much harder to believe that equal respect is truly the principle governing research decisions.

These subtleties did not, however, make their way into federal regulations governing research on human subjects.[7] Those regulations stipulated that the risks to fetuses *in utero* imposed by research must be "minimal," and that nonviable fetuses *ex utero* suffer "no added risk" as a result of research. Vital functions of the aborted fetus outside the uterus could not be artificially maintained for the sake of research, nor could the experiment itself terminate the heartbeat or respiration of the aborted (living, but nonviable) fetus outside the uterus.[8]

These guidelines governed research on pregnant women and their fetuses. It will, however, become important for the story we are tracing to note that the federal regulations also dealt with the possibility that human embryos might be created outside the womb *in vitro* and there become research subjects. The regulations stipulated that such research was not to be given federal funding without prior approval by an Ethics Advisory Board.[9] Because there was considerable opposition to funding such research, that advisory board existed only from 1978 to 1980. Because it was not replaced, no Ethics Advisory Board review could be carried out. Hence, research on embryos produced *in vitro* could not be carried out with federal funding.

A second round of debate, about a related issue, began in the late 1980s after reports of research in Mexico involving transplantation of fetal neural tissue into the brains of two patients suffering from Parkinson's disease. Because fetal cells grow quickly and divide rapidly, because they are less likely to be rejected at transplantation than adult tissues, and because they can be attained while they are still pluripotential and not yet differentiated into particular organ systems, their availability promised new possibilities in the treatment of a variety of disorders (Parkinson's, Alzheimer's disease, spinal cord damage, and diabetes, for example).

In 1988 the National Institutes of Health (NIH) appointed an ethics advisory committee to study the question, and a moratorium on federal funding of fetal tissue research was put in place until that committee completed its deliberations. Its report eventually favored (by a 17–4 majority) funding such experimentation, as long as certain guidelines were followed.[10] The most significant restrictions were that the mother of the fetus-to-be-aborted should not be permitted to know who would benefit from the aborted child's tissue, that no abortion and donation should be permitted between relatives, and that the possibility of tissue donation should be raised with a pregnant woman only *after* her decision for abortion had been made.

In this debate the issue of abortion clearly loomed large, and it is, in fact, difficult to disentangle the question of abortion from that of fetal tissue research. Perhaps the principal concern of those who opposed such research was that it appears to invite our cooperation in the great moral evil of abortion. To use for the benefit of others tissue from those we have chosen to kill seems to inflict a still further injustice on one already deprived of life. And the very fact that those who donate and use fetal tissue for research may often sincerely think of themselves as thereby bringing good out of evil is itself cause for concern. It constitutes a subtle invitation to self-deception, an invitation to ignore the reality in which we participate and which we choose.

We should not, however, overlook the fact that the issue being debated here is *not* the same as that taken up in the first round of argument. For this research did not propose to use any *living* fetuses as experimental subjects. It proposed only to use tissue from aborted, dead fetuses, having extracted from the tissue mass some cells for experimental transplantation. Even while granting this, we should note that some caution is in order here. Thus, for example, Mary Carrington Coutts writes: "In 1989, an article published by Swedish, British, and American authors described their method of retrieving fresh fetal brain tissue as suctioning brain tissue from the fetus while it was still alive *in utero*. . . . While this is believed to be an isolated incident in another country, the article raised some of the old concerns about the use of fetal tissue."[11] Moreover, even while the limited issue of research using tissue from dead (aborted) fetuses was ostensibly the subject of debate, other scholars were suggesting that federal regulations governing the use of still living but nonviable fetuses *ex utero* might be too stringent. Thus, Mahowald, Silver, and Ratcheson wrote: "We believe . . . that use of essential organs or tissue from nonviable [but still living] fetuses is morally defensible if dead fetuses are not available or are not conducive to successful transplants."[12] When such research had been prohibited in federal regulations, they noted, we had not yet known of new "[p]ossibilities for therapeutic use of fetal tissue. . . . If the clinical significance of current transplantation techniques had been anticipated, different solutions to the problem might have been proposed."[13] Perhaps. But against this we ought, however, to set the oft-cited but too seldom-heeded words of Hans Jonas:

"Let us not forget that progress is an optional goal, not an unconditional commitment, and that its tempo in particular, compulsive as it may become, has nothing sacred about it."[14]

Granting these complexities, it remains true that the explicit concern of this round of debate was the use of tissue from dead (aborted) fetuses for transplantation research. We can perhaps clarify our own judgments on the matter by two thought experiments that attempt to distinguish the several moral issues interwoven in debate. Would we object to research using tissue acquired *only* from spontaneously aborted (miscarried) fetuses? I think it is hard to argue that such research in itself would be wrong. We might also ask ourselves whether we would object to research using tissue acquired *only* from those abortions which, though induced and intended, were abortions we thought permissible (however small or large that class might be). This, at least in my view, is a harder call. To use for the benefit of others those whom we have already (even if legitimately) condemned to die is so clearly an example of the strong using the weak that we might well reject such research. As Kathleen Nolan has written: "The welfare of another being has been sacrificed [in abortion], however legitimately, for the good of society or someone else. A moral intuition insists that being used once is enough."[15] The restrictions on fetal tissue transplantation research suggested by the NIH advisory committee are probably sufficient to deal with worries about moral complicity in the encouragement or sanctioning of abortion. They are less able, however, to deal with Nolan's concern. Nevertheless, on January 21, 1993, President Clinton lifted the moratorium on fetal tissue research, stating in an executive order: "We must free science and medicine from the grasp of politics. . . ."[16]

The 1994 Report of the Human Embryo Research Panel constitutes yet a third round in this debate—though, again, the point at issue has its own particular nuances. The issue here is not experimental use of tissue taken from dead fetuses, nor is it experiments upon fetuses *in utero* or nonviable fetuses *ex utero.* Instead, the focus now is on the earliest stages of development of the embryo produced outside the womb *in vitro*—what the Report terms, erroneously I think, the "preimplantation human embryo."[17] I noted above the requirement in federal regulations that such research be funded only after review by an Ethics Advisory Board, which had been in existence only from 1978 to 1980. Because the Board had never been replaced, no such research could receive federal funding. When the U.S. Congress acted in 1993 to nullify the requirement that such research proposals be reviewed by the (nonexistent) Ethics Advisory Board, the door was opened for federal funding of human embryo research. Before providing funding, however, the National Institutes of Health appointed a panel to recommend guidelines to govern the research. To consider its report is to look at bioethics as it is often practiced today.

II

How should such a panel proceed? This panel clearly thought of itself as involved in the formulation of a neutral public policy. The basic statement of its perspective is, indeed, difficult to believe, so eager is it to eschew any perspective:

> Throughout its deliberations, the Panel considered the wide range of views held by American citizens on the moral status of preimplantation embryos. In recommending public policy, the Panel was not called upon to decide which of these views is correct. Rather, its task was to propose guidelines for preimplantation human embryo research that would be acceptable public policy based on reasoning that takes account of generally held public views regarding the beginning and development of

human life. The Panel weighed arguments for and against Federal funding of this research in light of the best available information and scientific knowledge and conducted its deliberations in terms that were independent of a particular religious or philosophical perspective. (P. 1)

The defects in such an approach need to be noted. Most important perhaps is that the decisions of such a panel are not likely to be as neutral as its depiction of its task would suggest. It is a little remarkable that, in an age when the Academy is awash in claims that we cannot detach ourselves from our limited, partial perspectives, panel members should so readily picture themselves as philosopher-kings who can adjudicate the disputes between conflicting views without themselves being parties to the conflict. One suspects that they could imagine this only if, from the start, they were already in agreement—only if, that is, the possibility of a true clash of viewpoints had been excluded in the constitution of the panel. Indeed, Ronald Green, a member of the panel, reporting on its deliberations, noted that some critics had objected to the makeup of the panel. "In reply . . . [NIH Director] Dr. Varmus and other NIH administrators pointed out that Panel members, who represented a broad spectrum of religious backgrounds, were selected primarily for their medical, scientific, or scholarly expertise and not on the basis of their positions on the acceptability of embryo research."[18] Green suggests that NIH might have avoided such criticism had it appointed one or more scholars or researchers known for their right-to-life views to the panel. "However," he writes, in a sentence one cannot help thinking he might like to have back, so breathtaking is it in its condescension, "this would have run counter to the primary focus on expertise."[19]

More important, however, for considering what the task of such a panel ought to be, is a second reason Green gives for thinking it might not have been good to appoint a panel of diverse views. "It also might have converted the open debate and discussion that marked the Panel's proceedings into a less productive matter of bargaining and negotiation between individuals with fixed positions."[20] One wonders why preordained agreement is preferable to bargaining and negotiation. Discussing the possibilities for moral agreement within a pluralistic society, Robert Merrihew Adams has written: "Compromise shares with majority vote the advantage that it does not require anyone to admit an error."[21] As Adams notes, neither compromise nor majority vote are forms of rational persuasion; they do not pretend that parties to a conflict have reached theoretical agreement. But they have proven remarkably successful as mechanisms that make possible life within a pluralistic society.

This suggests, in fact, that guidelines of the sort governing human embryo experimentation are better determined by political representatives, who cannot pretend to be above the fray in search only of the truth. Adams' reasoning is, again, compelling.

> Conflict is inevitable—and, I would add, not altogether undesirable. The meeting of will with will, which almost always involves conflict at some level, is the very substance of personal relationships. . . . It is through conflict, or something like it, that we know the otherness of self and other. . . . The fact that the world is in some ways contrary, and in some ways unresponsive, to our wills is what keeps us from regarding it all as an extension of ourselves. . . . In politics, likewise, conflict could hardly be eradicated without excluding from the political process the selfhood of most of the individuals, and the identity of many of the groups, in society.[22]

Especially when we are deliberating the fate of those whom some would regard as the weakest and least powerful *members* of our community, whose selfhood might easily be excluded and taken less seriously, the give and take of political argument is far preferable to the deliberations of a panel that supposes itself to be free of any religious or philosphical perspective. Responding precisely to such give and take, on the very same day that the advisory committee to the director of NIH voted to approve the report of the Human Embryo Research Panel, President Clinton banned the use of federal funds to support one important aspect of this research that had been approved by the panel: the creation of embryos specifically for purposes of research. Having recognized the difficult moral and political questions involved, and accepting explicitly that the research he was prohibiting could produce medical benefits, he stated: "However, I do not believe that federal funds should be used to support the creation of human embryos for research purposes, and I have directed that N.I.H. not allocate any resources for such research."[23] As I noted above, when two years earlier he had lifted the moratorium on fetal tissue transplantation research, President Clinton had said that science and medicine must be freed from politics. Here, in drastically altered political circumstances, he seemed to reassert that grasp. And, indeed, Patricia King, a co-chair of the Human Embryo Research Panel, was quoted as saying: "'I don't think that what the president did has as much to do with the merits of embryo research as politics.'"[24] She may well be right; however, what she perhaps meant as criticism I read more favorably. At issue in debates over human embryo research is a question of membership within our community. This is, if anything is, a political question, requiring a full clash between different religious and philosophical perspectives. The NIH panel would have served us better—and served the discipline of bioethics better—had it brought that clash much more fully into its report.

III

Although the panel report does not offer any detailed account of its method, it is not hard to see in its report the standard set of principles—beneficence and respect for autonomy, with an occasional nod toward the requirements of distributive justice.[25] And we have here a good example of how thin such a "principlist" approach can become when it lacks—or in this case deliberately eschews—a richer understanding of the human person. The panel begins with an acceptance of the status quo: *In vitro* fertilization is here to stay, but it is not as successful as it ought to be. More should be done to help infertile couples, and embryo research holds out the promise of such help. Beyond improving success with *in vitro* fertilization, a variety of other possibly beneficial results are foreseen: production of cell lines for use in tissue transplantation, new techniques for genetic screening, new contraceptive techniques. But, of course, over against such possible benefits we must set the principle of respect for persons. Hence, the report's third chapter, "Ethical Considerations in Preimplantation Embryo Research," discusses the moral status of the early embryo. Here it is hardly neutral among competing views.

The basic thrust of its discussion proceeds as follows: Two general views about the status of the embryo are distinguished. The first attempts to find some single criterion of moral personhood, some point in human development at which a person—with the full complement of equal rights—is present. But there are, it turns out, many different views about what this criterion should be, and they emphasize in turn quite different points in human development as the moment when personhood is present. The second view is characterized as pluralistic. It sees personhood more as a gradual development, not marked by the acquisition of any single characteristic.[26] Having described the several sorts of views to be found in current discussion and having discovered to no one's surprise that we are not in agreement on the status of the embryo, the panel reasserts its benign vision of its role: "Americans hold widely different views on the question of the moral value of prenatal life at its various stages. These views are often based on deeply held religious and ethical beliefs. It is not the role of those who help form public policy to decide which of these views is correct. Instead, public policy represents an effort to arrive at a reasonable accommodation to diverse interests" (p. 50).

Having reached this point, the panel might well have closed down its operation; in some sense, that is the conclusion to which its own statement leads. But, of course, it did not. Instead, it adopted a particular view of personhood, all the while denying that it was doing so and failing to adopt it in a thoroughgoing manner. In effect, it treats the formation of the primitive streak at fourteen days of development as decisive.[27] There is a good bit to be said for such a view, since before this time twinning may still occur. Hence, one can argue that only at fourteen days is individuation established and only then can we say that an individual human being (or more than one individual human being) is present. The panel was faced with several difficulties, however, that it did not address satisfactorily.

Current federal regulations governing fetal research do not make fourteen days of development decisive. They define the fetus simply as "the product of conception from the time of implantation . . . until a determination is made, following expulsion or extraction of the fetus, that it is viable."[28] They respect, we might say, the trajectory of the body, of human development even before individuality is decisively established. Current federal regulations (thrashed out in the first round of debate detailed above) also protect the embryo as potential research subject (even prior to fourteen days) when it is *in utero*. The implanted embryo may not be the subject of research that carries more than minimal risk. Those regulations also stipulate, as the panel noted, that "research on fetuses that are to be aborted and those that are to be carried to term be given equal treatment" (p. 52). Hence, our choice and intent ought not determine the degree of protection to which the embryo is entitled. Thus, in current regulations governing research upon human subjects, the ethics of experimentation are kept separate from the issue of abortion. If abortion is permissible, that is presumably because the interests of the fetus and the mother clash, and the mother's are given priority. When we contemplate research upon the embryo/fetus, however, no such clash is present, and therefore the fetus-to-be-aborted and the fetus-going-to-term must be given equal respect. Risks cannot be imposed upon the former if they would not be imposed upon the latter.

In order to make place for research upon the unimplanted embryo, however, the panel gave much more moral weight to human choice and will—and this in several ways. The fact that the *un*implanted embryo is regularly referred to as a *pre*-implantation embryo obscures the role of human choice. The panel regularly distinguishes between the protection required for embryos intended for transfer to a womb and embryos that will not be transferred—and it imposes far more rigorous standards of risk minimization when transfer is intended. "This distinction in treatment between embryos that will not be transferred and those that will be is warranted by the need to avoid harms to the child who will be born" (p. 52). Clearly, if we are to accept this distinction as morally significant, we must justify it in one of two ways: We might simply give up any claim to equal respect for all human embryos. Or, like the earlier National Commission report discussed above, we might argue that, although (formally) all embryos are respected equally, the material meaning of "minimal risk" changes dramatically when the embryo is one whose life we have no intention of preserving. In either case, the bestowal of equal respect or its meaning is at the mercy of our choices.[29]

Perhaps the clearest indication of how decisive human will has become in the panel's report is the approval it gives in certain circumstances to the deliberate creation of embryos for research purposes (an approval which, as we noted above, President Clinton overturned). Noting moral disagreements concerning such deliberate creation of embryos for research, the panel concludes, in a sentence that all too clearly suggests the poverty of bioethics as the search for public policy: "These arguments are metaphysically complex and controverted, and the Panel did not come to any conclusion about their validity or weight" (p. 54). A reader may perhaps be pardoned for wondering why the panel thinks it has come to no conclusion, since it immediately proceeds to permit embryo creation under certain circumstances because of the research benefits it promises. A discussion of those metaphysical complexities from such a panel might genuinely have enriched our public debate. Instead, it gives us a peculiar combination of metaphysical bewilderment and practical certitude. At the very least, we should be clear that at points such as this the panel's claim to philosophical neutrality is specious.

IV

Discussing the way in which doctors at Auschwitz were "hungry for surgical experience," Robert Jay Lifton writes: "In the absence of ethical restraint, one could arrange exactly the kind of surgical experience one sought, on exactly the appropriate kinds of 'cases' at exactly the time one wanted. If one felt Hippocratic twinges of conscience, one could usually reassure oneself that, since all of these people were condemned to death in any case, one was not really harming them."[30] Let us note that justification: Because, by virtue of decisions others had made, the victims had no future life prospects, they could not really be harmed if subjected to experiments that would never have been carried out on other people. If, in light of this chapter's discussion, that justification sounds familiar, we must simply face the fact. Structurally, it is the justification offered by some on the National Commission for the Protection of Human Subjects for contemplating what would ordinarily be considered research involving more than minimal risk on fetuses-to-be-aborted. Structurally, it is similar to the Human Embryo Research Panel's willingness to permit experiments involving greater risk of harm on embryos not intended for transfer.

To note such a point is, of course, immediately to anger those who do not wish to be compared with Nazi doctors, and the "Nazi analogy," commonly invoked in bioethical argument, is much disputed.[31] There is, I suspect, more to the analogy than is sometimes granted—particularly if we keep in mind the Nazi ideology's emphasis on personhood.[32] But my reason for appealing to it here is somewhat different. I do not claim that we may find ourselves on a slippery slope, at the bottom of which might lie Nazi-like acts. Nor do I intend to compare the substance of our moral choices to Nazi-like deeds. Instead, I want to recall how easily we may deceive ourselves about what we do, how subjectively well-meaning people may approve objective evil.

I have used Lifton's *The Nazi Doctors* with students on a number of occasions, and they never fail to find terrifying one of its central themes. They see that the book does not, in fact, invite us chiefly to ask whether some action we contemplate (such as embryo experimentation) might in fact be Nazi-like. It invites us to see within ourselves—good people that we suppose we are—the possibility of great evil. That is the terrifying power of Lifton's book. "We thus find ourselves returning," Lifton writes near the end of his discussion of the doubling that made it psychologically possible to be a Nazi doctor, "to the recognition that most of what Nazi doctors did would be within the potential capability—at least under certain conditions—of most doctors and of most people."[33] To read Lifton's account is, for the most part, to read of good and ordinary people in the grip of an ideology, who suppose themselves to be engaged in the purely scientific (and philosophically neutral) practice of medicine. And something like this, Lifton suggests, is an almost universal human possibility. Therefore, it seems to me, it is always fair, appropriate, and important to ask what kind of bioethics can best protect us against the possibilities for evil that may lie within us. I do not think we ought to place much confidence in a bioethics that, thinking itself free of religious or philosophical contamination, goes in search of public policy.

5. The Issue That Will Not Die

When we attempt to date the "birth of bioethics," we usually look a little earlier than January of 1973. And as a matter of historical explanation alone, we are probably right to do so. Nevertheless, the significance for bioethics of the *Roe v. Wade* decision cannot be overlooked. Perhaps more than any other single event, it has shaped our bioethical discourse. It has given autonomy—via the language of "privacy"—centrality, and it is hard to imagine that discussions about dying patients would have followed the course they have without that language. Indeed, we were led ineluctably to the vague and baffling language in the 1992 *Casey* decision about "the right to define one's own concept of existence, of meaning, of the universe, and of the mystery of human life."[1] That sentence did not gestate long before it bore fruit in a decision by a Federal judge in Seattle, who ruled that the suffering of a terminally ill person seeking suicide assistance is equally intimate and personal.[2] The *Roe v. Wade* decision was also of great significance in shaping the language of "personhood"—a language that has proved of incalculable importance in bioethical argument. Debates about both the beginning and the end of life have usually turned on the issues of autonomy and personhood. It is, therefore, worth attending to the nature of the abortion argument.

The central lines in that debate have remained remarkably stable over the last twenty years—so stable as to suggest an impasse. Two early and widely anthologized articles about abortion staked out these lines of argument with a clarity that in many ways remains unsurpassed. Mary Anne Warren's "On the Moral and Legal Status of Abortion" argued that fetuses, lacking personhood, were not entitled to protection against being killed.[3] I will call this the "personhood" argument. Judith Jarvis Thomson's essay, "A Defense of Abortion," articulated in unforgettable fashion the case for holding that, even were the personhood of the fetus established, women would not be obligated to use their bodies to support fetal life.[4] I will call this the "bodily support" argument. In my judgment, neither proves to be very satisfactory without the other, but to commit ourselves to the claims of both may require a more expansive faith than we can or should summon.

I

The argument about the moral status—the personhood—of the fetus arises quite naturally and perhaps inevitably. After all, the claim that "it's my body to do with as I wish" becomes more worrisome if that body is nourishing another human life equal in dignity. These two lives may be for a time inseparably joined, but if the fetus can claim a dignity like ours, we will surely have to worry about aborting it. Thus has the language of personhood entered the abortion debate—first as a legal term in *Roe*, but then in moral argument more generally.

It is important to note that personhood language, in the context of this debate, generally serves to exclude rather than include. The irony of our public debate is that the people labeled "conservative" are often those arguing for a more liberal and inclusive understanding of the meaning of human personhood. This language once served to include within our common humanity those whose skin color was not white. Now it increasingly serves as a language by which we mark off those human subjects who *cannot* lay claim to equal protection. We do well to ponder the implications of that fact. As Philip Abbott has written:

> There are very few general laws of social science but we can offer one that has a deserved claim: the restriction of the concept of humanity in any sphere never enhances a respect for human life. It did not enhance the rights of slaves, prisoners of wars, criminals, traitors, women, children, Jews, blacks, heretics, workers, capitalists, Slavs, Gypsies. The restriction of the concept of personhood in regard to the fetus will not do so either.[5]

At issue here is not simply an important moral debate, but our understanding of the humanity we share.

If we do not wish to reduce our humanity to either its material or its spiritual dimension, we need a concept of the person that does justice to the duality of our nature. But when we attempt to avoid the now dreaded Cartesian image of the "ghost in the machine," we often turn to an entirely functional understanding of the human being: A person is simply a body capable of intentional, self-aware action. And then what do we make of the "person" who, though he may be *terra animata*, lacks such capacities? Even if there are satisfactory ways to handle the old question whether a person sleeping is still a person, it will prove more difficult on this view to ascribe personhood to fetuses, infants and young children, and the senile.

Thus, for example, in an important work that gathers together many strands of this increasingly common view, H. Tristram Engelhardt asserts that persons are those who can be "concerned about moral arguments and . . . convinced by them. They must be self-conscious, rational, free to choose, and must possess moral concern."[6] It is clear that, judging on the basis of these criteria, many human beings will not qualify for personhood. And our obligations to them will be less stringent than to those—like ourselves—who are self-conscious, rational, and in control. Engelhardt does not hesitate, for example, to speak of parents as *owning* their children, at least until such time as the children become self-conscious. Just as arguments about the Nazi analogy never end in bioethics, so also we may argue at length whether it is helpful to use analogies from the history of our nation's struggle over slavery when thinking about abortion, but surely this language of ownership is striking. It is the almost certain result of restrictive and exclusivistic thinking about the meaning of personhood.

At stake here is the development and enlargement of our concept of human community. Perhaps the ability to sustain a bond of affection that unites us across generations is as fundamental as "self-awareness" for our understanding of what it means to be human.[7] Moreover, there are deeper theoretical difficulties with a narrow and exclusivistic understanding of personhood. For there is a difference between the characteristics that distinguish the human species, and the qualifications for membership in the species. It may be that among the distinguishing characteristics of humans are features such as rationality and self-consciousness. But one can be human without exercising (or even having the capacity to exercise) such characteristics. To be human one need only be begotten of human parents. Indeed, those who lack some of these capacities are best described as the weakest and least advantaged *members* of our community. The fetus should be cared for and protected not because of any "personal" capacities, but because weak and vulnerable human beings—who, lacking some of our qualities, do not lack equality with us—are the weakest members of the human community.

This does not mean that abortion is never permissible, but it does make clear how heavy a burden the "bodily support" argument, grounded in autonomy, must carry. It will have to justify killing one who is equal in dignity to each of us. Granting, then, that the fetus is genuinely one of us, need a woman provide it with her bodily support?

II

The practical assertion that, since pregnancy involves a woman's body, the choice of continuing that pregnancy must be hers alone, was first given powerful theoretical articulation and defense by Judith Jarvis Thomson. Although the personhood argument is also involved in the *Roe v. Wade* decision, the bodily support argument has perhaps played an even larger role in the moral argument and debate surrounding abortion.

Stripped to its essentials, the argument asserts: We are embodied beings; thus the person is involved when the body is given or used.[8] To require women to continue an unwanted pregnancy is, therefore, to ask of them personal sacrifice. Since men cannot become pregnant (and be required to make such sacrifice), prohibiting abortion is, in effect, institutionalizing sexual inequality.

All three elements in the argument are necessary, though they are not always consistently developed. Thomson, for example, did not do full justice to the sense in which we *are* our bodies. She argued that it would never be unjust of a woman to abort, though there might be circumstances in which it would be indecent for her to do so. But then, discussing pregnancy resulting from rape, Thomson engaged in one too many thought experiments: She contemplated the possibility of abortion if the pregnancy were to last only an hour rather than nine months. In that case, Thomson held, although the woman would not act unjustly in aborting the fetus, it would be indecent of her not to carry it to term. Yet surely this misses the reason why pregnancy resulting from rape constitutes a special case. Because the woman's *body* has been used and violated, her *person* has been assaulted—and this would be just as true if the pregnancy came to term within an hour. The argument needs a richer sense of our embodied personhood than Thomson herself offered.

That richer sense has been supplied by Patricia Beattie Jung, who argues that we should think of childbearing and organ donation as analogous activities.[9] Each offers a kind of bodily support, and each therefore involves a certain element of personal sacrifice. Since men have not been required to serve as organ donors even when (because of tissue type) they could well do so, Jung suggests that required childbearing does indeed mandate sexual inequality. A similar point has been argued by Laurence Tribe, who notes that "current law nowhere forces *men* to sacrifice their bodies and restructure their lives even in those tragic situations (of needed organ transplants, for example) where nothing less will permit their children to survive. . . ."[10]

We may well, however, have doubts about the validity of the analogy. After all, current law also nowhere forces a *woman* to serve as an organ donor. Perhaps what the law would reflect if abortion were regulated is not institutionalized sexual inequality but some sense of an important difference between organ donation and abortion. When a man (or a woman) declines to serve as organ donor, and when we in turn decline to compel him or her to do so, what does *not* happen might be termed a rescue operation. But potential donors, even if they are not required to rescue the imperiled person in need of an organ, are not permitted to aim at that person's death. That I decline to make the bodily and personal sacrifice of giving you my kidney does not entitle me to asphyxiate you, nor does it entitle me to stop others who might wish to offer you a kidney. If aborting a fetus meant only ceasing to carry it while permitting others to sustain its life—which, of course, it cannot mean, at least for the present—the analogy might seem more persuasive. Declining to donate a kidney and aborting a fetus may both be actions that *result* in death, but they differ in the important sense that only the latter can be said to aim at death. If the day comes when it is medically possible to stop carrying a fetus without at the same time aiming at its death, we will be able to test the validity of the analogy more carefully in our actual practice.

One might, of course, redescribe abortion in a way that makes the analogy seem more plausible. Tribe proposes a thought experiment: If women automatically miscarried after conception unless they took a drug to prevent such a miscarriage, a law compelling them to take it would mandate a rescue operation calling for bodily sacrifice. Declining to take the drug, then, would not be aiming to kill the fetus; it would only be failing to rescue. And since, unlike required organ donation, such a law could apply only to women, it would seem to institutionalize sexual inequality.

What is striking about such a thought experiment is the way in which it invites us to alienate ourselves from the natural manner in which conception and birth take place.[11] We are invited to picture the fetus as an alien who may perhaps be invited in but who, if not welcomed, is essentially an invader. In pregnancy resulting from forcible intercourse, and perhaps also in other circumstances in which the woman's consent to intercourse has been greatly reduced, the thought experiment may well have merit. But that there are such hard cases should not obscure the more general truth: that a fetus is not demanding access to some place it has no right to be; that it is simply seeking to survive in its natural environment. I do not suppose, of course, that every fetus in its natural environment will at every moment be wanted or desired by its mother. But as Sidney Callahan has written, "morality also consists of the good and worthy acceptance of the unexpected events that life presents."[12] And *that* is a principle that can and should be applied equally to fathers and mothers. Rather than adopting a position that "ratifies the view that pregnancies and children are a woman's private responsibility," we should do all in our power to encourage and require parental responsibility.[13] Because a fetus is unwanted, it does not become an alien who may be destroyed. To justify that conclusion, the "bodily support" argument would need considerable assistance from the "personhood" argument. Each feeds unwittingly off the other; yet neither is adequate.

This makes clear why abortion is an issue that will not—and should not—die. We can defend the current status quo only by buying deeply into the concepts of personhood and autonomy as they have developed over the past several decades. I have tried to show in earlier chapters that these concepts are problematic and that our debates about them are far from resolved. Nor will any minimalist understanding of public policy be adequate to this issue; for we are exploring the most basic of political questions: the outer limits of our community and the meaning of membership in it. If *Roe v. Wade* was in part responsible for the rise to prominence of the concepts of personhood and autonomy, perhaps reflection upon their difficulties will bring us full circle to a reconsideration of *Roe v. Wade* itself. That would not be bad for bioethics—or for the rest of us.

Notes

CHAPTER I

1. Albert R. Jonsen, "The Birth of Bioethics," *Hastings Center Report* 23 (November/December 1993). Special Supplement, S1. Alexander's article, "They Decide Who Lives, Who Dies," appeared in the November 1962 issue of *Life.*

2. Albert R. Jonsen, "American Moralism and the Origin of Bioethics in the United States," *Journal of Medicine and Philosophy* 16 (1991): 114.

3. David J. Rothman, *Strangers at the Bedside: A History of How Law and Bioethics Transformed Medical Decision Making* (New York: Basic Books, 1991), 3. The Beecher article appeared in *New England Journal of Medicine* 74 (1966): 1354–60.

4. Larry R. Churchill, "Reviving a Distinctive Medical Ethic," *Hastings Center Report* 19 (May/June 1989): 30.

5. Daniel Callahan, "Morality and Contemporary Culture: The President's Commission and Beyond," *Cardozo Law Review* 6 (1984): 348.

6. Leon Kass, "Neither for Love nor Money: Why Doctors Must Not Kill," *The Public Interest* 94 (1989): 25–46.

7. Paul F. Camenisch, *Grounding Professional Ethics in a Pluralistic Society* (New York: Haven Publications, 1983), 9.

8. Kass, "Why Doctors Must Not Kill," 41. In addition, see Ezekiel J. Emanuel, *The Ends of Human Life: Medical Ethics in a*

Liberal Polity (Cambridge and London: Harvard University Press, 1991), 16–22.

9. Churchill, "Medical Ethic," 29. Our attitudes toward medical paternalism are often paradoxical, however. On the one hand, there has been great concern about what Jay Katz termed "the silent world of doctor and patient," the failure of doctors really to invite patients into decision making. But, on the other hand, concerns about cost containment have led us to turn physicians into gatekeepers who must be less than patient centered as they allow larger socioeconomic considerations to have an impact upon their treatment decisions. Surely it is ironic that physicians are becoming the servants of institutions (both public and private) at the same time that business managers are aspiring to the status of "professionals."

10. Callahan, "Morality and Contemporary Culture," 348.

11. Ibid.

12. Robert M. Veatch, "Medical Ethics: Professional or Universal?" *Harvard Theological Review* 65 (1972): 559.

13. Churchill, "Medical Ethic," 33.

14. Emanuel, *Ends of Human Life*, 23.

15. Ibid., 171.

16. Robert Merrihew Adams, "Religious Ethics in a Pluralistic Society," in *Prospects for a Common Morality*, ed. Gene Outka and John P. Reeder, Jr. (Princeton, N.J.: Princeton University Press, 1993), 105. I assume the need for some agreement in constitutional law on matters that cannot simply be settled by votes. Nevertheless, as I note below, if all my interests (or your interests) are given absolute constitutional protection, few alternatives to violence will be left for the resolution of disagreements.

17. My governing assumption about human nature was given well known formulation by Reinhold Niebuhr: "Man's capacity for justice makes democracy possible; but man's inclination to injustice makes democracy necessary." See *The Children of Light and the Children of Darkness* (New York: Charles Scribner's Sons, 1944), xiii.

18. Adams, "Religious Ethics," 106.

19. Cf. Richard L. Fern, "Religious Belief in a Rawlsian Society," *Journal of Religious Ethics* 15 (spring 1987): 33–58.

20. Emanuel, *End of Human Life*, 32.

21. Callahan, "Morality and Contemporary Culture," 350.

22. Tom Beauchamp and James Childress, *Principles of Biomedical Ethics* (New York and Oxford: Oxford University Press, 1989). Here I use the third edition. The fourth edition has now been issued, though I was not able to use it during the writing of this chapter.

23. Edwin R. DuBose, Ronald P. Hamel, and Laurence J. O'Connell, eds., *A Matter of Principles?: Ferment in U.S. Bioethics* (Valley Forge, Pa.: Trinity Press International, 1994).

24. Beauchamp and Childress, *Principles of Biomedical Ethics*, 394. Further references to this book will be cited by page number in parentheses within the body of the text.

25. John P. Reeder, Jr., "Foundations Without Foundationalism," in Outka and Reeder, *Prospects for a Common Morality*, 191–214. Beauchamp and Childress do not defend this view of morality at any length. They simply characterize their position as nonfoundationalist in footnote 20, p. 24.

26. Cf., for example, Outka and Reeder, *Prospects for a Common Morality*, 193.

27. Childress has recently made clear that the approach is not intended to be as deductive as it may seem. "Although PBE [*Principles of Biomedical Ethics*] has used the metaphor of *application*, as in *applied ethics*, this was a mistake that misled some readers, as did its charts that sometimes appeared to indicate a top-down approach through a process of justification." Cf. DuBose, Hamel, and O'Connell, *A Matter of Principles?*, 81.

28. The National Commission for the Protection of Human Subjects of Biomedical and Behavioral Research, *The Belmont Report: Ethical Principles and Guidelines for the Protection of Human Subjects of Research* (Washington, D.C.: U.S. Government Printing Office, 1978).

29. The procedure must inevitably be described as a form of intuition. In *A Matter of Principles?* (p. 71) Childress describes it as follows: "A great deal rests on what has been variously called prudence, practical moral reasoning, or discernment in the situation. . . ." Later (p. 81) he says that their method is probably best described as a "balancing, which depends on intuitive weighing of conflicting principles." Of course, the more we describe the method in such terms,

the more difficult it may be to distinguish it from one of its chief competitors—casuistry—which I take up below.

30. Cf. Gilbert Meilaender, "Is What is Right for Me Right for All Persons Similarly Situated?" *Journal of Religious Ethics* 8 (spring 1980): 125–134.

31. Cf. William F. May, "Attitudes Toward the Newly Dead," *Hastings Center Studies* 1 (1973): 3–13; and Leon R. Kass, M.D., "Thinking About the Body," chapter 11 in *Toward a More Natural Science* (New York: The Free Press, 1985).

32. DuBose, Hamel, and O'Connell, *A Matter of Principles?*, 88.

33. Courtney S. Campbell, "Principlism and Religion: The Law and the Prophets," in DuBose, Hamel, and O'Connell, *A Matter of Principles?*, 195.

34. It is important to note that Beauchamp and Childress are clearer than many others have been about the meaning of their commitment to autonomy. They distinguish (p. 373) between the ideal of the autonomous person and the principle of respect for personal autonomy. They are committed to the latter, not the former. And, in fact, they question whether we ought even share the ideal of the autonomous person. Still, their failure to provide a more substantive account of the "self" presupposed by their principle of respect for autonomy or "self"-determination means that the ideal of the autonomous person (as one who chooses his life plan and acts independently of any external authority) is likely to govern most uses of their theory.

35. Campbell, "Principalism and Religion," 196f.

36. Cf. Albert R. Jonsen and Stephen Toulmin, *The Abuse of Casuistry: A History of Moral Reasoning* (Berkeley, Los Angeles, London: University of California Press, 1988).

37. Ibid., 7.

38. Albert R. Jonsen, "On Being a Casuist," in *Clinical Medical Ethics: Exploration and Assessment*, ed. Terrence F. Ackerman, Glenn C. Graber, Charles H. Reynolds, and David H. Thomasma (Lanham, Md.: University Press of America, 1987), 123.

39. Jonsen, "American Moralism," 123.

40. Robert M. Veatch, "From Forgoing Life Support to Aid-in-Dying," *Hastings Center Report* 23 (November/December 1993). Special Supplement on "The Birth of Bioethics," S7.

41. Daniel Callahan, "Why America Accepted Bioethics," *Hastings Center Report* 23 (November/December 1993). Special Supplement on "The Birth of Bioethics," S8.

42. Jonsen, "American Moralism," 120.

43. Ibid., 119.

44. Ibid., 126. This reading of the history requires us, of course, to view the *Belmont Report*, published by the National Commission for the Protection of Human Subjects of Biomedical and Behavioral Research—the first of our two national commissions—as an aberration. For this report, in which the "principlist" position was clearly outlined, was, on Jonsen's account, authored by a commission in which the casuists were gradually overcoming the moralists. We cannot, of course, discount the firsthand quality of Jonsen's testimony, since he was a commissioner on that body and also served later on the President's Commission for the Study of Ethical Problems in Medicine. He now describes the *Belmont Report* as "a product of American moralism, prompted by the desire of Congressmen and of the public to see the chaotic world of biomedical research reduced to order by clear and unambiguous principles" (Ibid., 125). The Commission's actual work, he argues, was essentially casuistic in character. So dubious is the rest of his historical analysis, however, that no one need rush to accept this judgment. Moreover, his vision of "American moralism" is so intense that it makes him almost incapable of recognizing serious moral challenge, since he reads the psyches of those who disagree as unable to deal with moral complexity. Thus, for example, he discusses the treatment by the President's Commission of providing life support for patients in a persistent vegetative state. The Commission held that there was no moral obligation to provide such support, a conclusion that has been challenged at least in the instance of artificial nutrition and hydration. "Yet," Jonsen writes, "the ethical reasoning developed to justify the original proposition remains unchallenged in concept and in logic. Ethical argument has encountered American moralism" (Ibid., 126). The complacence of such a statement lends substance to a suspicion that the revived casuistry gives rise to a very tame and complacent ethic.

45. Ibid., 127.

46. Ibid.

47. Stephen Toulmin, *The Place of Reason in Ethics* (Chicago and London: University of Chicago Press, 1986).

48. Ibid., 28.

49. Ibid., xix. I doubt whether Toulmin can really eschew principles as much as he thinks; for the distinction between principles and cases can never be as sharp as he makes it. Even to reason analogously from case to case begins to imply the recognition of an inchoate principle in accord with which the cases are rightly connected.

50. Of course, this is not the only issue. As I noted earlier, Beauchamp and Childress develop their principlist approach while themselves renouncing foundationalism in moral theory.

51. *Republic*, 7.514a–517c.

52. Iris Murdoch, *The Sovereignty of Good* (London: Routledge and Kegan Paul, 1970), 84.

53. Richard Rorty, *Philosophy and the Mirror of Nature* (Princeton, N.J.: Princeton University Press, 1979), 376.

54. Stephen Toulmin, "The Tyranny of Principles," *Hastings Center Report* 11 (December 1981): 37.

55. Ibid., 34.

56. Ibid., 37.

57. Toulmin, *The Place of Reason in Ethics*, 142.

58. Jonsen and Toulmin, *The Abuse of Casuistry*, 7.

59. Ibid., 45.

60. Ibid., 66.

61. Ibid., 46.

62. For what follows see also Eric T. Juengst, "Casuistry and the locus of certainty in ethics," *Medical Humanities Review* 3 (January 1989), 19–27.

63. Jonsen and Toulmin, *The Abuse of Casuistry*, 285. See also p. 323, where three types of "problematic" cases are distinguished. The third and most perplexing is described as "those occasional—but unavoidable—circumstances in which our basic moral categories are called into question."

64. Ibid., 317.

65. Juengst, "Casuistry," 26.

66. DuBose, Hamel, and O'Connell, *A Matter of Principles?*, 90.

67. Ibid., 91.

68. Jonsen and Toulmin, *The Abuse of Casuistry*, 338.

69. Philip Turner, review of *The Abuse of Casuistry*, by Jonsen and Toulmin, *Religious Studies Review* 17 (October 1991), 305.

70. Daniel Callahan, "Why America Accepted Bioethics," S9.

71. Cf. Henry S. Richardson, "Specifying Norms as a Way to Resolve Concrete Ethical Problems," *Philosophy and Public Affairs* 19 (fall 1990), 279–310; and David De Grazia, "Moving Forward in Bioethical Theory: Theories, Cases, and Specified Principlism," *Journal of Medicine and Philosophy* 17 (October 1992), 511–39.

72. Richardson, "Specifying Norms," 286.

73. Ibid., 283.

74. Paul Ramsey, "The Case of the Curious Exception," in *Norm and Context in Christian Ethics*, ed. Gene H. Outka and Paul Ramsey (New York: Charles Scribner's Sons, 1968), 67–135. For an illuminating description of Ramsey's casuistry as one that "surrounds the particular," see D. Stephen Long, "Protestant Casuist," chapter 3 in *Tragedy, Tradition, Transformism: The Ethics of Paul Ramsey* (Boulder, Colo., San Francisco, Oxford: Westview Press, 1993).

75. Ibid., 87.

76. Ibid., 88.

77. Paul Ramsey, *The Patient as Person: Explorations in Medical Ethics* (New Haven and London: Yale University Press, 1970), 113–64.

78. Ibid., 157–64.

79. Paul Ramsey, *Ethics at the Edges of Life: Medical and Legal Intersections* (New Haven and London: Yale University Press, 1978), 223.

80. "The Self-Giving of Vital Organs: A Case Study in Comparative Ethics," chapter 4 in *The Patient as Person*.

81. Ibid., 195.

82. Ibid.

83. Shana Alexander, "Thirty Years Ago," *Hastings Center Report* 23 (November/December 1993). Special Supplement on "The Birth of Bioethics," S5.

84. Robert S. Morison, "Bioethics After Two Decades," *Hastings Center Report* 11 (April 1981), 8.

85. Callahan, "Why America Accepted Bioethics," S8.

86. Ibid.

87. James P. Wind, "What Can Religion Offer Bioethics?" *Hastings Center Report* 20 (July/August 1990). Special Supplement, 19.

88. Courtney S. Campbell, "Religion and Moral Meaning in Bioethics," *Hastings Center Report* 20 (July/August 1990). Special Supplement, 9.

89. Leon R. Kass, "Practicing Ethics: Where's the Action?" *Hastings Center Report* 20 (January/February 1990), 5–12.

90. Ibid., 8.

91. Ibid.

92. Ibid., 11.

93. Ibid., 6.

94. Ibid., 12.

CHAPTER 2

1. Saint Augustine, *The City of God*, trans. Henry Bettenson (New York: Penguin Books, 1972), 20.20. Future citations will be given by book and chapter number within parentheses in the body of the text.

2. For much of what follows about the early Fathers I draw upon J. N. D. Kelly, *Early Christian Doctrines* (New York, Evanston, Ill., and London: Harper and Row, 1960), 464–79. I am indebted to Robert Wilken for drawing my attention to Kelly's discussion.

3. C. S. Lewis, *Miracles* (New York: Macmillan, 1947), 166.

4. Saint Thomas Aquinas, *Summa Contra Gentiles*, trans. Charles J. O'Neil (Notre Dame and London: University of Notre Dame Press, 1975), 4.81.12. Future citations will be given by book, chapter, and paragraph number within parentheses in the body of the text.

5. Jonathan Edwards, *Ethical Writings*, ed. Paul Ramsey, vol. 8 of *The Works of Jonathan Edwards*, ed. John E. Smith (New Haven and London: Yale University Press, 1989), 371.

6. Austin Farrer, *Love Almighty and Ills Unlimited* (Garden City, N. Y.: Doubleday and Company, 1961), 166. For his discussion more generally, see the appendix, "Imperfect Lives," 166–68.

7. David H. Smith, *Health and Medicine in the Anglican Tradition* (New York: Crossroad, 1986), 10.

8. Holmes Rolston III, "The Irreversibly Comatose: Respect for the Subhuman in Human Life," *Journal of Medicine and Philosophy* 7 (1982): 342.

9. Robertson Davies, *The Rebel Angels* (New York: Penguin Books, 1983), 249f.

10. Joseph Fletcher, "Indicators of Humanhood: A Tentative Profile of Man," *Hastings Center Report* 2 (November 1972): 1–4.

11. Joseph Fletcher, "Four Indicators of Humanhood—The Enquiry Matures," *Hastings Center Report* 4 (December 1974): 4–7.

12. I have discussed this from another angle in chapter 8 of *Faith and Faithfulness* (Notre Dame, Ind.: University of Notre Dame Press, 1991).

13. Lawrence J. Schneiderman, Nancy S. Jecker, and Albert R. Jonsen, "Medical Futility: Its Meaning and Ethical Implications," *Annals of Internal Medicine* 112 (June 15, 1990): 950.

14. Ronald E. Cranford, "The Persistent Vegetative State: The Medical Reality (Getting the Facts Straight)," *Hastings Center Report* 18 (February/March 1988): 28.

15. Schneiderman, Jecker, and Jonsen, "Medical Futility," 952.

16. Bart Collopy, Philip Boyle, and Bruce Jennings, "New Directions in Nursing Home Care," *Hasting Center Report* 21 (March/April 1991). Special Supplement 1–16.

17. Paul Ramsey, *The Patient as Person* (New Haven and London: Yale University Press, 1970), xiii.

18. Hans Jonas, "The Burden and Blessing of Mortality," *Hastings Center Report* 22 (January/February 1992): 35.

19. For the historical information that follows I rely upon George J. Annas, "The Health Care Proxy and the Living Will," *New England Journal of Medicine* 324 (April 25, 1991): 1210–13.

20. Alexander Morgan Capron, "In Re Helga Wanglie," *Hastings Center Report* 21 (September/October 1991): 26–28.

21. My distinction here bears some similarities to James Childress' distinction between autonomy as an end state and autonomy as a side constraint. Cf. his *Who Should Decide? Paternalism in Health Care* (New York and Oxford: Oxford University Press, 1982), 64.

22. Schneiderman, Jecker, and Jonsen, "Medical Futility," 952.

23. John A. Robertson, "Second Thoughts on Living Wills," *Hastings Center Report* 21 (November/December 1991): 6–9.

24. Rebecca Dresser and Peter J. Whitehouse, "The Incompetent Patient on the Slippery Slope," *Hastings Center Report* 24 (July/August 1994): 7.

25. Ibid., 11.

26. I have discussed this point more fully (and acknowledged my indebtedness for it to Oliver O'Donovan) in *Faith and Faithfulness*, 45–47.

27. Leon R. Kass, *Toward A More Natural Science* (New York: The Free Press, 1985), 293.

28. James Rachels, *Created From Animals: The Moral Implications of Darwinism* (Oxford and New York: Oxford University Press, 1990), 198ff.

29. How widespread such a view has become may be seen from the fact that even Dresser and Whitehouse ("The Incompetent Patient on the Slippery Slope"), with their very sensitive treatment of demented persons, adopt the view that "consciousness is a prerequisite for patients to benefit directly from, or to have an interest in, continued life" (p. 7). Even patients who are "'barely conscious,'" with "'negligible awareness of self, other, and the world,'" may not, in their view, "retain any direct interest in having their lives continued" (p. 10).

30. John Kleinig, *Valuing Life* (Princeton, N.J.: Princeton University Press, 1991), 201.

31. Rolston, "Irreversibly Comatose," 352.

32. Ibid., 338.

CHAPTER 3

1. John A. Robertson, *Children of Choice: Freedom and the New Reproductive Technologies* (Princeton, N.J.: Princeton University Press, 1994). Citations throughout this chapter will be given by page number in parentheses within the body of the text.

2. Cf. Paul Ramsey, *Fabricated Man: The Ethics of Genetic Control* (New Haven and London: Yale University Press, 1970); and

Joseph Fletcher, *The Ethics of Genetic Control: Ending Reproductive Roulette* (Garden City, N.Y.: Anchor, 1974).

3. For a similar argument, cf. Paul Lauritzen, *Pursuing Parenthood: Ethics Issues in Assisted Reproduction* (Bloomington and Indianapolis, Ind.: Indiana University Press, 1993), 42.

4. I discuss these two types of arguments in more detail in chapter 5.

5. Cf. Lauritzen, *Pursuing Parenthood*, 56–63.

6. C.S. Lewis, *A Preface to Paradise Lost* (London, Oxford, New York: Oxford University Press, 1942), 102.

7. Oliver O'Donovan, *Begotten Or Made?* (Oxford: Clarendon Press, 1984), 16. The discussion that follows draws upon O'Donovan's argument at pages 16–17.

8. Ibid, 17.

9. Ibid.

10. Paul Ramsey, *Fabricated Man*, 36.

11. Cf. *Nicomachean Ethics*, 6.2–5.

12. H. Richard Niebuhr, *The Responsible Self: An Essay in Christian Moral Philosophy* (New York, Evanston, Ill., and London: Harper and Row, 1963). My use of Niebuhr's categories here has been shaped and influenced by Thomas W. Ogletree, "Values, Obligations, and Virtues: Approaches to Bio-medical Ethics," *Journal of Religious Ethics* 4 (1976): 105–30.

13. Ogletree, "Values, Obligations, and Virtues," 114.

14. Niebuhr, *The Responsible Self*, 108f.

15. O'Donovan's point in choosing a crucial phrase from the Nicene Creed ("begotten, not made") as the title of his book was precisely to engage us in exploring the meaning of the dignity of the person.

16. Gabriel Marcel, "The Mystery of the Family," in *Homo Viator: Introduction to a Metaphysic of Hope* (New York: Harper Torchbooks, 1962), 88.

17. Ibid., 90.

18. Ibid., 87.

19. Paul Tillich, *Dynamics of Faith* (New York: Harper and Brothers Publishers, 1957), 42.

20. Paul Ricoeur, *The Symbolism of Evil* (Boston: Beacon Press, 1967), 348.

21. Ibid., 9.

22. Cf. my "The Fetus as Parasite and Mushroom: Judith Jarvis Thomson's Defense of Abortion," *Linacre Quarterly* 46 (May 1979): 126–35.

23. C.S. Lewis, "Transposition," in *The Weight of Glory and Other Addresses* (Grand Rapids, Mich.: Eerdmans, 1949), 28.

24. O'Donovan, *Begotten Or Made?* 86.

CHAPTER 4

1. Although it has not been published, the Report was formally presented on September 27, 1994. This report was subject to review by the Advisory Committee to the Director. But on the same day that the Advisory Committee endorsed the Report's guidelines, President Clinton announced that his administration would prohibit federal funding for *creating* human embryos solely for research purposes— thus overriding one central recommendation in the Report. Citations of the Report will be given by page numbers in parentheses within the body of the text.

2. National Commission for the Protection of Human Subjects of Biomedical and Behavioral Research, *Research on the Fetus: Report and Recommendations* (Washington, D.C.: Department of Health, Education, and Welfare, 1975).

3. See recommendations 1, 4, and 7 of the Commission's report, *Research on the Fetus*.

4. See recommendations 5 and 6 of the Commission's report. It is important to note that each of these recommendations contains an "escape clause" calling for a national ethical review body to consider "research presenting special problems related to the interpretation or application of these guidelines. . . ." The "special problems" involve the view of some Commissioners that fetuses-to-be-aborted could be subjected to research risks that should not be imposed on fetuses-going-to-term.

5. *Research on the Fetus*, 66f.

6. "Dissenting Statement of Commissioner David W. Louisell," in *Research on the Fetus*, 77.

7. See the Code of Federal Regulations (CFR) governing Protection of Human Subjects (45 CFR 46).

8. 45 CFR 46:208–9.

9. Cf. 45 CFR 46:204d: "No application or proposal involving human *in vitro* fertilization may be funded by the Department or any component thereof until the application or proposal has been reviewed by the Ethical Advisory Board and the Board has rendered advice as to its acceptability from an ethical standpoint."

10. *Report of the Human Fetal Tissue Transplantation Research Panel* (Bethesda, Md.: Department of Health and Human Services, National Institutes of Health, December 1988).

11. Mary Carrington Coutts, "Fetal Tissue Research," *Kennedy Institute of Ethics Journal* 3 (March 1993): 82–83.

12. Mary B. Mahowald, Jerry Silver, and Robert A. Ratcheson, "The Ethical Options in Transplanting Fetal Tissue," *Hastings Center Report* 17 (February 1987): 12.

13. Ibid., 11.

14. Hans Jonas, "Philosophical Reflections on Experimenting with Human Subjects," in *Readings on Ethical and Social Issues in Biomedicine*, ed. Richard W. Wertz (Englewood Cliffs, N.J.: Prentice Hall, 1973), 35.

15. Kathleen Nolan, "*Genug ist Genug*: A Fetus Is Not a Kidney," *Hastings Center Report* 18 (December 1988): 14.

16. Stephen Burd, "U.S. Won't Back Creation of Human Embryos for Research," *Chronicle of Higher Education*, 14 December 1994, A32.

17. Such a formulation obscures the crucial role of human choice and will here. Cf. "The Inhuman Use of Human Beings: A Statement on Embryo Research by the Ramsey Colloquium," *First Things* (January 1995): 19: "In question are not preimplantation embryos but *unimplanted* embryos—embryos produced with the intention that they will *not* be implanted. . . ."

18. Ronald M. Green, "At the Vortex of Controversey: Developing Guidelines for Human Embryo Research," *Kennedy Institute of Ethics Journal* 4 (December 1994): 347.

19. Ibid.

20. Ibid.

21. Robert Merrihew Adams, "Religious Ethics in a Pluralistic Society," in *Prospects for a Common Morality*, ed. Gene Outka and John P. Reeder, Jr. (Princeton, N.J.: Princeton University Press, 1993), 104.

22. Ibid., 106.

23. Stephen Burd, "Creation of Human Embryos for Research," A32.

24. John Schwartz and Ann Devroy, "President limits funds for research on embryos," *Cleveland Plain Dealer*, 3 December 1994, 1-A.

25. The nod toward distributive justice takes the following form: "There must be equitable selection of donors of gametes and embryos and efforts must be made to ensure benefits and risks are fairly distributed among subgroups of the population" (p. 79).

26. The panel report (p. 49, n. 13) says that Ronald Green, a member of the panel, had provided a philosophical statement of the pluralistic approach in an article titled, "Toward a Copernican Revolution in Our Thinking About Life's Beginning and Life's End," *Soundings* 66 (Summer 1983): 152–73. I doubt that this article is as revolutionary as Green suggests, but, in any case, I cannot find in it any extended defense of the "pluralistic" approach of the panel's report. The bulk of Green's article is devoted not to arguing that there is no single point at which personhood becomes present, but instead to arguing that such a point must be determined by those of us who are already indisputably persons, who are adult moral agents. It is not simply "given" in the nature of things. We must, in his view, "balance" against any set of characteristics that candidates for personhood possess the restrictions on our own liberty that granting them personhood will involve. Out of that balancing comes a decision about granting or withholding of moral personhood and its attendant rights. But when Green writes that we should not "fixate on a set of qualities possessed or not possessed by the entity and lose sight of the complex relational balancing judgment we must make" (p. 167), he is not describing what the panel report calls a pluralistic understanding of personhood. He is not balancing a number of developmental con-

siderations in order to trace a gradually developing claim to equal treatment. He is balancing the claims of the candidate for personhood—claims that might be based on some single characteristic—against our own broadest interests and the limits we will suffer if we grant the claim to personhood. This is precisely how he himself describes his thesis: "Above all, my aim has been to show that the questions of the beginning and the end of life require collective human moral decisions, ones that are inescapably ours to make. What I have been opposing is the view that the answers to these questions are, in some way, *out there* merely waiting to be discovered" (p. 172). This is not at all the panel's concern, which is best described as a conflict between two different views of personhood, each dependent on characteristics "out there"—one pointing to some single characteristic as decisive, the other to some developing ensemble of characteristics. Only quite near the end of his article does Green move toward something closer to the panel's pluralistic approach, and even then a good bit of ambiguity remains. He writes:

> To this point and throughout my remarks I have suggested that personhood and protectability are all or nothing matters, that a being is or is not a person, is or is not fully protectable. In one sense I believe this is true. If personhood can be taken to mean equal membership in the human moral community and the corresponding inviolability of a being's life, then one either is or is not a person. One cannot be half a person. Yet even if we reserve the term 'person' for those who possess this full inviolability, we might wish to limit our liberty and to accord at least a measure of moral respect to beings that are not 'persons' in this sense but which our broadest human interests counsel protecting. (P. 168)

This, I guess, comes somewhat closer to the kind of view the panel had in mind, but it hardly constitutes the main line of argument in Green's article. Moreover, the main line of argument—that the bestowal of personhood is a decision to be made by those who are already indisputably persons—runs quite counter to the report's perhaps naive belief that it is simply examining the facts free of any philosophical captivity.

27. This view is characterized as "a compromise among competing viewpoints" necessary "for public policy purposes" (p. 60). An exception is made for "research protocols with the goal of reliably identifying in the laboratory the primitive streak" (p. 79). And the limitation is clearly described as one that ought to be in place "for the present" (p. 79).

28. 45 CFR 46:203c.

29. In this respect, at least, the panel's report was faithful to the thrust of (panel member) Ronald Green's argument. Cf. Green, "Toward a Copernican Revolution," 161: "The particular judgment that a class of beings is protectable results when we as adult human beings decide that a set of qualities in that kind of being is sufficiently important in terms of our broadest human interests to merit the restraint on our liberty that an acknowledgment of personhood involves."

30. Robert Jay Lifton, *The Nazi Doctors* (New York: Basic Books, 1986), 295. Cf. also Edward Zukowski, "The 'Good Conscience' of Nazi Doctors," in *Annual of the Society of Christian Ethics: 1994*, 58, where Zukowski describes part of the defense at Nuremberg offered by Karl Gebhardt: "He argued that since the people he experimented upon were already condemned to death, he thought it best that 'some good be brought out of evil.'"

31. Cf., for example, the commentaries by Nat Hentoff, Daniel Callahan, Gary E. Crum, and Cynthia B. Cohen in "Contested Terrain: The Nazi Analogy in Bioethics," *Hastings Center Report* 18 (August/September 1988): 29–33. In "At the Vortex of Controversy," Ronald Green notes that "Panel members were frequently likened to Hitler or Stalin and accused of complicity in genocide. (An accusation of this sort, made during a public presentation, provoked an angry and emotional response from Chairman Steven Muller, himself a refugee from Nazi persecution)" (p. 348). But, of course, no one is more dangerous than the person whose conscience is clear and confident.

32. Nazi experiments began, as is well known, with "impaired" children and adults. That something like a "pluralistic" understanding of personhood was at work here is hard to doubt. Cf., for example, Lifton's description of the views of Ernst Haeckel (pp. 441f.).

33. Lifton, *The Nazi Doctors*, 427.

CHAPTER 5

1. *Planned Parenthood Association of Southeastern Pennsylvania v. Robert P. Casey,* 60 *United States Law Weekly* 4800 (1992).

2. Timothy Egan, "U.S. Judge Says Constitution Protects Right to Suicide Aid," *New York Times,* 5 May 1994, A1.

3. *Monist,* 57 (January 1973): 43–61. Warren later added a "Postscript on Infanticide." The article with postscript can be found in *Ethical Issues in Modern Medicine,* ed. Robert Hunt and John Arras (Palo Alto, Calif.: Mayfield, 1977), 159–77.

4. *Philosophy and Public Affairs* 1 (fall 1971): 47–66.

5. Philip Abbott, "Philosophers and the Abortion Question," *Political Theory* 6 (August 1978): 329.

6. H. Tristram Engelhardt, Jr., *The Foundations of Bioethics* (New York: Oxford University Press, 1986), 105.

7. Abbott, "Philosophers and the Abortion Question," 332.

8. On this point the "bodily support" argument sees the truth of our nature far more clearly than the "personhood" argument, which generally loses any sense of the person as "animated earth," as living body.

9. Patricia Beattie Jung, "Abortion and Organ Donation: Christian Reflections on Bodily Life Support," *Journal of Religious Ethics* 16 (fall 1988): 273–305.

10. Laurence A. Tribe, *Constitutional Choices* (Cambridge, Mass.: Harvard University Press, 1985), 243–44.

11. Thus, for example, Tribe can elsewhere write that "equality for women must mean the same ability to express human sexuality without the burden of pregnancy and childbirth that has always been, *by accident of biology,* available to men." *Abortion: The Clash of Absolutes* (New York and London: W.W. Norton, 1990), 212, emphasis added. The divorce of person and body could not be more clearly stated.

12. Sidney Callahan, "Abortion and the Sexual Agenda," *Commonweal* 25 (April 1986): 235.

13. Ibid., 236.

Index